Student Study Guide
Volume I
for
Tipler and Mosca's
Physics for Scientists and Engineers
Sixth Edition

Todd Ruskell
Colorado School of Mines

W.H. Freeman and Company
New York

Acknowledgments

I must first thank Paul Tipler and Gene Mosca for putting together an excellent textbook in the sixth edition of *Physics for Scientists and Engineers*. It has been a delight to work with. I must also thank Gene for his work on earlier versions of this study guide, which have been drawn on heavily.

I am indebted to the reviewers of this study guide: Elizabeth Behrman (Wichita State), Daniel Dale (University of Wyoming), Linnea Hess (Olympic College), Nikos Kalogerpoulos (City University of New York), and Oren Quist (South Dakota State). Their careful reading and insightful comments made this a much better study guide than it would have been otherwise. In addition, their thorough reviews helped uncover many errors that would have remained to be discovered by the users of this study guide. Their assistance is greatly appreciated. In spite of our combined best efforts, there may still be an occasional error in this study guide, and for those I assume full responsibility. Should you find errors or would like to bring another matter regarding this study guide to my attention, please do not hesitate to send them to me by using asktipler@whfreeman.com.

My wife Susan and daughter Allison provided immense support with their patience and understanding throughout the entire process of writing this study guide.

It was a pleasure to work with Susan Brennan, Clancy Marshall, Kharissia Pettus, and Kathryn Treadway who guided me through the creation of this study guide.

April, 2007

Todd Ruskell
Colorado School of Mines

Contents

1 Measurement and Vectors   1

PART I   MECHANICS   17

2 Motion in One Dimension   17

3 Motion in Two and Three Dimensions   33

4 Newton's Laws   47

5 Additional Applications of Newton's Laws   71

6 Work and Kinetic Energy   91

7 Conservation of Energy   113

8 Conservation of Linear Momentum   131

9 Rotation   145

10 Angular Momentum   171

R Special Relativity   187

11 Gravity   197

12 Static Equilibrium and Elasticity   217

13 Fluids   229

PART II   OSCILLATIONS AND WAVES   243

14 Oscillations   243

15 Traveling Waves   259

16 Superposition and Standing Waves   277

PART III   THERMODYNAMICS   291

17 Temperature and Kinetic Theory of Gases   291

18 Heat and the First Law of Thermodynamics   305

19 The Second Law of Thermodynamics   329

20 Thermal Properties and Processes   347

Answers to Problems   357

To the Student

This study guide was written to help you master Chapters 1 through 20 of Paul Tipler and Eugene Mosca's *Physics for Scientists and Engineers*, Sixth Edition. Each chapter of the study guide is divided into sections that match the textbook, and culminates in a short quiz designed to test your mastery of the subject. Each of these sections may contain the subsections below.

In a Nutshell: A brief overview of the important concepts presented in the section. This section is designed only to remind you of the key ideas. If a concept is not clear, you should refer back to the text for more detailed explanations and derivations.

Physical Quantities and Their Units: A list of the constants, units, and physical quantities introduced in the section.

Fundamental Equations: A list of fundamental equations introduced in the section. These expressions provide underpinning for the Important Derived Results.

Important Derived Results: In many sections the Fundamental Equations are applied to specific physical situations. This application can result in important derived results that apply to only those specific situations. These results are listed here.

Common Pitfalls: Warnings about commonly-made mistakes. In addition, there are conceptual questions designed to test your understanding of the physical principles discussed in the section.

Try It Yourself: Workbook style questions following the structure of the solutions to the worked Examples in the text. Final answers, with units, are provided, but it is up to you to use the space provided to fill in the required work for the intermediate steps. Most of these questions also have a Taking It Further question designed to enhance your understanding of and ability to interpret the problem's solution. You should answer these questions in the space provided before looking at the answers in the back of the study guide.

What Is the Best Way to Study Physics?

Of course there isn't a single answer to that. It is clear, however, that you should begin early in the course to develop the methods that work best for you. The important thing is to find the system that is most comfortable and effective for you, and then stick to it.

In this course you will be introduced to numerous concepts. It is important that you take the time to be sure you understand each of them. You will have mastered a concept when you fully understand its relationships with other concepts. Some concepts will seem to contradict other concepts or even your observations of the physical world. Many of the questions in this study guide are intended to test your understanding of concepts. If you find that your understanding of an idea is incomplete, don't give up; pursue it until it becomes clear. We recommend that you keep a list of the things that you come across in your studies that you do not understand. Then, when you come to understand an idea, remove it from your list. After you complete your study of each chapter, bring your list to your most important resource, your physics instructor, and ask for assistance. If you go to your instructor with a few well-defined questions, you will very likely be able to remove any remaining items from your list.

Like the Example problems presented in the textbook, the problem solutions presented in this study guide start with basic concepts, not with formulas. We encourage you to follow this practice. Physics is a collection of interrelated basic concepts, not a seemingly infinite list of disconnected, highly specific formulas. Although at times it may seem we present long lists of formulas, do not try to memorize long lists of specific formulas and then use these formulas as the starting point for solving problems. Instead, focus on the concepts first and be sure that you understand the ideas before you apply the formulas.

Probably the most rewarding (but challenging) aspect of studying physics is learning how to apply the fundamental concepts to specific problems. At some point you are likely to think, "I understand the theory, but I just can't do the problems." If you can't do the problems, however, you probably don't understand the theory. Until the physical concepts and the mathematical equations become your tools to apply at will to specific physical situations, you haven't really learned them. There are two major aspects involved in learning to solve problems: drill and skill. By drill we mean going through a lot of problems that involve the direct application of a particular concept until you start to feel familiar with the way it applies to physical situations. Each chapter of the text contains about 35 single-concept problems for you to use as drill. Do a lot of these—at least as many as you need in order to feel comfortable handling them.

By skill we mean the ability both to recognize which concepts are involved in more advanced, multi-concept problems, and to apply those concepts to particular situations. The text has several intermediate-level and advanced-level problems that go beyond the direct application of a single concept. As you develop this skill you will master the material and become empowered. As you find that you can deal with more complex problems—even some of the advanced-level ones—you will gain confidence and enjoy applying your new skills. The examples in the textbook and the problems in this study guide are designed to provide you with a pathway from the single-concept to the intermediate-level and advanced-level problems.

A typical physics problem describes a physical situation—such as a child swinging on a swing—and asks related questions. For example: If the speed of the child is 5.0 m/s at the bottom of her arc, what is the maximum height the child will reach? Solving such problems requires you to apply the concepts of physics to the physical situation, to generate mathematical relations, and to solve for the desired quantities. The problems presented here and in your textbook are exemplars; that is, they are examples that deserve imitation. When you master the methodology presented in the worked-out examples, you should be able to solve problems about a wide variety of physical situations.

A good way to test your understanding of a specific solution is to take a sheet of paper, and—without looking at the worked-out solution of an Example problem—reproduce it. If you get stuck and need to refer to the presented solution, do so. But then take a fresh sheet of paper, start from the beginning, and reproduce the entire solution. This may seem tedious at first, but it does pay off.

This is not to suggest that you reproduce solutions by rote memorization, but that you reproduce them by drawing on your understanding of the relationships involved. By reproducing a solution in its entirety, you will verify for yourself that you have mastered a particular example problem. As you repeat this process with other examples, you will build your very own personal base of physics knowledge, a base of knowledge relating occurrences in the world around you-the physical universe-and the concepts of physics. The more complete the knowledge base that you build, the more success you will have in physics.

You should budget time to study physics on a regular, preferably daily, basis. Plan your study schedule with your course schedule in mind. One benefit of this approach is that when you study on a regular basis, more information is likely to be transferred to your long-term memory than when you are obliged to cram. Another benefit of studying on a regular basis is that you will get much more from lectures. Because you will have already studied some of the material presented, the lectures will seem more relevant to you. In fact, you should try to familiarize yourself with each chapter before it is covered in class. An effective way to do this is first to read the In a Nutshell subsections of that study guide chapter. Then thumb through the textbook chapter, reading the headings and examining the illustrations. By orienting yourself to a topic before it is covered in class, you will have created a receptive environment for encoding and storing in your memory the material you will be learning.

Another way to enhance your learning is to explain something to a fellow student. It is well known that the best way to learn something is to teach it. That is because in attempting to articulate a concept or procedure, you must first arrange the relevant ideas in a logical sequence. In addition, a dialog with another person may help you to consider things from a different perspective. After you have studied a section of a chapter, discuss the material with another student and see if you can explain what you have learned.

Chapter 1

Measurement and Vectors

1.1   The Nature of Physics

In a Nutshell

In ancient Greece, Aristotle's system of natural philosophy was based on assumptions, not actual experimentation. The Italian Galileo Galilei began the era of experimental science, driving our knowledge of the natural world, or "physics."

"Classical physics," the knowledge required to understand the macroscopic phenomena of our everyday world, includes mechanical motion, heat, sound, light, electricity, and magnetism. The first five parts of the text investigate primarily classical physics.

Part VI of the text introduces the realm of "modern physics"— relativity, radioactivity, wave-particle duality, and quantum theory. The application of these ideas to very small—atomic and subatomic—structures, and very fast—near light speed—phenomena has lead to a completely new level of understanding our everyday world.

1.2   Units

In a Nutshell

Physical quantities are numbers obtained by measuring physical phenomena. Some examples include the mass of this book, the time it takes you to sneeze, the thickness of a hair, and etc.

Units are standard quantities used to compare the relative sizes of physical quantities. For example, a book that is 2 cm thick is not as thick as a book that is 3 cm thick. However that same 2-cm book is thicker than a 2-mm thick book. The number by itself is meaningless. We need to include the unit used in that measurement in order to properly evaluate the size of the physical quantity being measured.

This text uses primarily the **International System** of units, or **SI** units. You may also occasionally hear this system referred to as the **mks** or meter-kilogram-second system. This system has seven fundamental units.

Time is measured in seconds, abbreviated s. The second is now defined as the period of time equal to 9 192 631 770 cycles of a specific energy transition in a cesium atom.

Length or distance, is measured in meters, abbreviated m. The meter is defined as the distance light travels in 1/(299 792 458) seconds.

Mass is measured in kilograms, abbreviated kg. The kilogram is now defined as the mass of a specific platinum-iridium alloy cylinder that is kept in Sèvres, France.

The other four fundamental units are the kelvin (K), the ampere (A), the mole (mol), and the candela (cd). These will be discussed in more detail later in the text.

When the names of units are written out, they always begin with a lowercase letter, *even when* the unit is named for a person. Thus, the unit of temperature named for Lord Kelvin is the kelvin. The abbreviation for a unit also begins with a lowercase letter, *except when* the unit is named for a person. Thus, the abbreviation for the meter is m, but the abbreviation for kelvin is K. The exception is the abbreviation for the liter, which is L. Abbreviations of units are not italicized and are not followed by periods unless at the end of a sentence.

When dealing with very large or very small numbers, it can be very useful to take advantage of unit prefixes, which are a shortcut for indicating powers of 10. All written-out prefixes, such as mega for 10^6, begin with lowercase letters. For multiples less than or equal to 10^3, the abbreviations of the prefixes are lowercase letters, such as k for kilo. For multiples greater than 10^3, the prefixes abbreviations are uppercase letters, such as M for mega. The prefixes for powers of 10 and their abbreviations are listed in Table 1-1 on page 5 of the text.

Other systems of units also exist. One common system is the cgs system, based on the centimeter, gram, and second. In the United States, a unit of force, the pound, is a fundamental unit. The fundamental unit of length is the foot, and the fundamental unit of time is still the second.

Physical Quantities and Their Units

Length or distance	meter (m), foot (ft), yard (yd)
Distance conversion factors	$1 \text{ ft} = \frac{1}{3} \text{ yd} = 0.3048 \text{ m}$
Time	second (s)
Mass	kilogram (kg)

Common Pitfalls

> ➤ It is important to memorize the powers of 10 that correspond to the prefixes, especially the most commonly used prefixes from nano to giga. See Table 1-1 on page 5 of the text.
> ➤ Remember that in the SI system, the kilogram is the fundamental unit of mass, and that $1 \text{ kg} = 10^3 \text{ g}$.
> ➤ Be careful when entering numbers using "e-notation" in your calculator. In this notation, the value 10 is written as 1e1, not 10e1.
> ➤ Always include units every time you write down a quantity. Consistently doing so provides an extremely useful check of your calculations.

1. TRUE or FALSE: The length of the meter depends on the duration of a second.

2. Is it possible to define a system of units for measuring physical quantities in which one of the fundamental units is not a unit of length? Why or why not?

Try It Yourself #1

Write the following in scientific notation without prefixes: (a) 330 km, (b) 33.7 μm, (c) 0.03 K, (d) 77.5 GW.

Picture: Each of the prefixes corresponds to multiplication by a factor of ten. Multiply each given number by that factor, and then rewrite the number with only one digit to the left of the decimal point.

Solve:

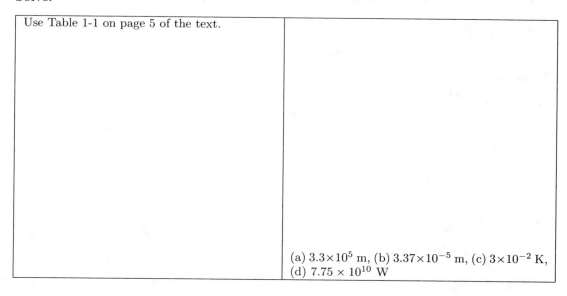

Use Table 1-1 on page 5 of the text.

(a) 3.3×10^5 m, (b) 3.37×10^{-5} m, (c) 3×10^{-2} K, (d) 7.75×10^{10} W

Try It Yourself #2

Write the following with prefixes, without using scientific notation: (a) 3.45×10^{-4} s, (b) 2.00×10^{-11} W, (c) 2.337×10^8 m, (d) 6.54×10^4 g.

Picture: When using prefixes, you should have only 1, 2, or 3 digits to the left of the decimal point. If there would be more than that, use the next larger prefix. If the digit to the left of the decimal point would be zero, use the next smaller prefix.

Solve:

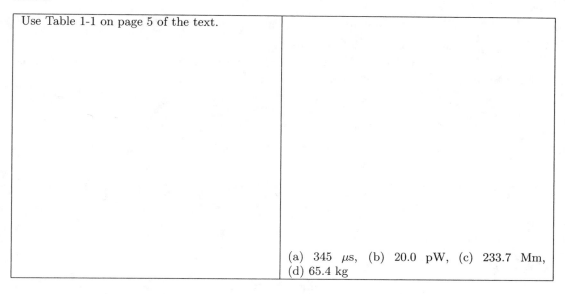

Use Table 1-1 on page 5 of the text.

(a) 345 μs, (b) 20.0 pW, (c) 233.7 Mm, (d) 65.4 kg

1.3   Conversion of Units

In a Nutshell

When physical quantities are added, subtracted, multiplied, or divided in an algebraic equation, the unit can be treated like any other algebraic quantity. For example, like units can be factored out or canceled as shown.

$$\frac{23 \text{ s}}{1} \times \frac{1 \text{ min}}{60 \text{ s}} \times \frac{1 \text{ h}}{60 \text{ min}} = \frac{23}{3600}\text{h}$$

$$23 \text{ m} + 42 \text{ m} = (23 + 42) \text{ m} = 65 \text{ m}$$

One unit can be converted to another by multiplying it by a **conversion factor** of magnitude 1, whose units depend on the conversion being made. Here we show the conversion factor between the SI unit of length, the meter, and the foot, a unit of length common in the United States. By writing out the units explicitly and canceling them as shown, you do not need to think about whether you multiply by 0.3048 or divide by 0.3048 to change meters to feet, because the units tell you whether you have chosen the correct or incorrect factor.

1 ft ≈ 0.3048 m. Therefore,

$$1 = \frac{0.3048 \text{ m}}{1 \text{ ft}} = \frac{1 \text{ ft}}{0.3048 \text{ m}}$$

See the front pages of the textbook or Appendix A of the textbook for a variety of useful conversion factors.

Common Pitfalls

> ➤ Usually, it is best to use SI units. If you are given problems with quantities in other units, you should usually convert them to SI units before proceeding with the problem.
> ➤ When converting units, remember that all conversion factors must have a magnitude of 1.

3. TRUE or FALSE: All conversion factors are equal to 1 and are unitless.

4. A furlong is one-eighth of a mile and a fortnight is two weeks. A furlong per fortnight is a unit of speed. What is the SI unit for this quantity? Find the conversion factor between furlongs per fortnight and the corresponding SI unit.

Try It Yourself #3

A very fast sprinter can run the 100-m dash in slightly under 10.0 s. This means his average speed is slightly greater than 10.0 m/s. Using conversion factors, convert 10.0 m/s to miles per hour.

Picture: We need a conversion factor for meters and miles, and one for seconds and hours.

Solve:

Find the conversion factor between seconds and hours.	
Find the conversion factor between meters and miles. There are approximately 1.61 km in one mile.	
Multiply the speed by conversion factors until you are left with miles per hour.	22.4 mi/h

Check: This speed falls into the range of "everyday" speeds that we experience.

Taking It Further: If you are not used to working in SI units, it can be difficult to gauge the feasibility of some quantities like speed, so conversion factors like 1 m/s = 2.24 mi/h can be useful to remember when trying to put your results into context. Write down at least five other units of speed.

Try It Yourself #4

The speed of light in empty space is 3×10^8 m/s. Use conversion factors to determine the speed of light in feet per nanosecond.

Picture: Two conversion factors will be needed, one for meters and feet, and another for seconds and nanoseconds.

Solve:

Find the conversion factor between feet and meters.	
Find the conversion factor between seconds and nanoseconds.	
Multiply by the conversion factors to find the speed in ft/ns.	0.984 ft/ns

Check: Make sure all the other units cancel out in your unit conversion. Often it is hard to assess whether or not a given numerical answer makes sense. In these situations, you should at least be able to verify that your units coincide with the quantity you are trying to calculate.

Taking It Further: Using only units analysis, how would you use this speed to determine how far a beam of light travels in 15 ns?

1.4   Dimensions of Physical Quantities

In a Nutshell

The **dimensions** of a physical quantity express what kind of quantity it is—whether it is a length, a time, a mass, or some other physical quantity. For example, the dimensions of velocity are length per unit time, which we write as $[v] = L/T$. The corresponding units might be miles per hour or meters per second. Whenever we add or subtract quantities, they must have the same dimensions. Also, both sides and every complete term in each side of an equation must have the same dimensions. The dimensions of some common physical quantities are shown in Table 1-2 on page 8 of the text.

Common Pitfalls

> ➤ *Always* make sure that both sides of an equation have the same dimensions. This kind of **dimensional analysis** can be a very efficient way to find mistakes in setting up your equations.

5. TRUE or FALSE: All conversion factors are equal to 1 and are dimensionless.

6. If two quantities are to be added, do they have to have the same dimensions? The same units? What if they are to be divided? Explain.

Try It Yourself #5

In the following equation, the distance x is in meters, the time t is in seconds, and the velocity v is in meters per second. What are the dimensions and SI units of the constants A, B, C, D, and E?

$$x = Avt + B\sin(Ct) + Dt^{1/2}3^{x/E}$$

Picture: Exponents, trigonometric functions, and the arguments of trigonometric functions are dimensionless. (The argument of $\cos\phi$ is ϕ). Quantities that are added must have the same dimensions.

Solve:

Exponents must be dimensionless. Use this to find the dimensions and units of E.	$[E] = L$, in meters
The argument of the sin function must be dimensionless. Use this to find the dimensions and units of C.	$[C] = 1/T$, with units 1/s

Each term on the right-hand side must have the dimensions of x, which is a length. Use this to find the dimensions and units of B.	
	$[B] = \text{L}$, in meters
Each term on the right-hand side must have the dimensions of x, which is a length. Use this to find the dimensions and units of A.	
	A is dimensionless, so it is also unitless.
Each term on the right-hand side must have the dimensions of x, which is a length. Use this to find the dimensions and units of D.	
	$[D] = \text{L}/\text{T}^{1/2}$, with units of $\text{m}/\text{s}^{1/2}$

Try It Yourself #6

In the following equation the distance x is in meters, the time t is in seconds, and the velocity v is in meters per second. What are the dimensions and SI units of the constants A, B, C, D, and E?

$$v = Bt\left[\sqrt{Ax} + \cos^2(Ct)\right] - D2^{Et}$$

Picture: Exponents and the arguments of trigonometric functions must be dimensionless. Trigonometric functions are dimensionless. Quantities that are added must have the same dimensions.

Solve:

Exponents and arguments of trigonometric functions are dimensionless, so the dimensions and units of C and E must both cancel the dimensions and units of t.	
	$[C] = [E] = 1/\text{T}$, in units of s^{-1}

The trigonometric function itself is dimensionless. That means $\sqrt{Ax}$ must also be dimensionless. Find the dimensions and units of A.	$[A] = 1/\text{L}$, with units of $1/\text{m}$
Since the quantity in square brackets is unitless, the product Bt must have the same dimensions and units as v. Use this observation to find the dimensions and units of B.	$[B] = \text{L}/\text{T}^2$, with units of m/s^2
The number 2 is unitless, but the product $D2^{Et}$ must have the same dimensions and units as v. Use this observation to find the dimensions and units of D.	$[D] = \text{L}/\text{T}$, with units of m/s

1.5 Significant Figures and Order of Magnitude

In a Nutshell

A **significant figure** is a reliably known digit (other than a zero that locates a decimal point).

When multiplying or dividing quantities, the number of significant figures in the final answer is no greater than that in the quantity with the fewest significant figures.

When adding or subtracting quantities, the number of decimal places in the answer should match that of the term with the smallest number of decimal places.

If the numbers involved in a calculation are very large or very small, you may want to rewrite these numbers in scientific notation. This notation often makes it easier for you to determine the number of significant figures that a number has and makes it easier for you to perform calculations.

In doing rough calculations, we sometimes round off a number to the nearest power of 10. Such a number is called an **order of magnitude**. Order-of-magnitude estimation calculations can be quite useful in answering questions that appear to have not enough information.

Common Pitfalls

> This course will require you to solve a large number of problems. For most of these, you will be using a calculator. When using very large or very small numbers, make sure you use scientific notation. If you do not, your calculator may automatically round off numbers inappropriately (unbeknownst to you), and your resulting calculations could be in error.

> When determining the number of significant figures, be careful of numbers like 1200. This could have anywhere from 2 to 4 significant figures. In this Study Guide, you can assume that all zeros are significant, but this is not a general rule that you can apply at any time.

> Order-of-magnitude estimates are also extremely helpful in determining the approximate correctness of a calculation. For instance, you know that if you multiply a number between 10 and 100, by another number between 10 and 100, you should get something between 100 and 10 000. If your calculator reports a number outside this range, you should instantly realize you made a mistake in entering the numbers into your calculator. Perhaps you put a decimal point in the wrong place, or entered too many or too few zeros.

7. TRUE or FALSE: In scientific notation, $10^0 = 10$.

8. If you use a calculator to divide 3411 by 62.0, you will get something like 55.016 129. (Exactly what you will get depends on your calculator.) Of course, you know that not all these figures are significant. How should you write the answer?

Try It Yourself #7

Use rough approximations to estimate the number of steps you would have to take to walk from Walworth, WI, to Annapolis, MD.

Picture: How far is it between these two cities? How big is a typical step?

Solve:

Estimate the distance between the two cities. (You may want to look this up on the Internet).	
Estimate the length of your typical step.	
Estimate the total number of required steps.	1.5 million

Check: Knowing that there are approximately 5280 ft in one mile, determine the distance, in miles, your number of steps represents. Compare this to the approximate coast-to-coast distance across the United States. Your answer should be something less than half that total distance.

Taking It Further: If you could walk nonstop at a constant pace, how long would it take you to walk this distance?

1.6 Vectors

In a Nutshell

Quantities with magnitude but no associated direction, such as speed, mass, volume, and time, are called **scalars**.
Quantities that have magnitude and direction, such as velocity, acceleration, and force, are called **vectors**. Graphically, we represent vectors by arrows that point in the direction of the physical quantity. The length of the vectors drawn to scale represents the magnitude, or size, of the vector quantity. 
In printed materials vectors are frequently represented by boldface symbols, for example **A** and **B**. Both here and in the text, they will be represented by placing an arrow over a boldfaced, italicized symbol; for example, $\vec{A}$ and $\vec{v}$. The magnitude of $\vec{A}$ is written $|\vec{A}|$, $\|\vec{A}\|$, or simply A. The magnitude of a vector is *never* negative.

Common Pitfalls

➢ The variable A could represent a purely scalar quantity, or it could be the magnitude of the vector $\vec{A}$. You need always to be aware of the context in which the variable is being used so that you can determine if it is *just* a scalar, or a scalar that is also the magnitude of a vector.

1.7 General Properties of Vectors

In a Nutshell

If an object moves from an initial position to a final position, we can represent its **displacement** with an arrow pointing from the initial to the final positions, as shown. The length and direction of this arrow are the magnitude and direction of the displacement vector. The displacement vector depends only on the initial and final positions of the object, not on the actual path the object follows. 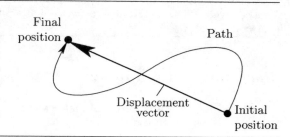

If an object moves from point 1 to point 2, and then from point 2 to point 3, the net displacement is the same as if the object moved directly from point 1 to point 3. The displacement $\vec{A}$ from point 1 to point 2 followed by the displacement $\vec{B}$ from point 2 to point 3 is equivalent to displacement $\vec{C}$ from point 1 to point 3. That is, $\vec{A} + \vec{B} = \vec{C}$. To add vectors graphically, draw the arrow $\vec{A}$. Then draw the arrow representing $\vec{B}$ with its tail at the head of the arrow representing $\vec{A}$. Then draw the arrow representing $\vec{C}$ from the tail of $\vec{A}$ to the head of $\vec{B}$. The arrows must all be drawn to scale, and the direction of the arrows is the same as the direction of the actual displacement. $\vec{C}$ is the **vector sum** or **resultant**.

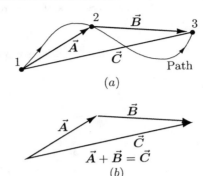

Vector subtraction, in which $\vec{D} = \vec{A} - \vec{B}$, is nothing more than adding the vector $-\vec{B}$ to $\vec{A}$. The opposite of a vector has the same magnitude as the original vector, but points in the opposite direction.

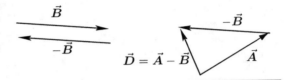

When a vector is multiplied by a positive scalar, the resulting vector points in the same direction, but its magnitude is altered by a factor equal to the scalar.

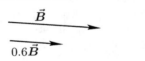

It is often useful to break a vector into its **components**. Consider $\vec{A}$. A_x, the x component of the vector, is the projection of the vector along the x axis. Similarly, A_y, the y component, is the projection of the vector along the y axis. For the geometry shown, we have the following relationships:

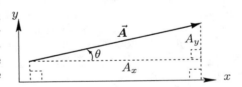

$$A_x = A\cos\theta \qquad A = \sqrt{A_x^2 + A_y^2}$$
$$A_y = A\sin\theta \qquad \tan\theta = A_y/A_x$$

Unit vectors are dimensionless vectors with a magnitude of 1. Their sole purpose is to point in a particular direction. Unit vectors are denoted by a "hat" above the variable. To find a unit vector that points in the direction of a specific vector, we divide that vector by its magnitude: $\hat{A} = \vec{A}/A$.

The unit vectors $\hat{\imath}$, $\hat{\jmath}$ and $\hat{k}$ are commonly used to indicate the positive x, y, and z directions, respectively. Thus a vector can be written as a function of its components and these unit vectors: $\vec{A} = A_x\hat{\imath} + A_y\hat{\jmath} + A_z\hat{k}$.

Important Derived Results

Component-vector relations, if $\vec{A}$ makes an angle θ with respect to the $+x$ axis

$$A_x = A\cos\theta \qquad A = \sqrt{A_x^2 + A_y^2}$$
$$A_y = A\sin\theta \qquad \tan\theta = A_y/A_x$$

Common Pitfalls

> ➤ Remember to use the appropriate notation to indicate vectors and unit vectors. Without this notation you will not be able to keep your variables straight.
> ➤ When you add vectors, remember that you cannot simply add their magnitudes. The equation $\vec{C} = \vec{A} + \vec{B}$ does not mean that $C = A + B$. Both these equations are true only if $\vec{A}$ and $\vec{B}$ are in the exact same direction.
> ➤ Be careful when using the arctan function to determine the angle a vector makes with the x axis. Your calculator will *always* return a value between $\pm 90°$. You need to look at the directions of the components of the vector in order to place the vector in the proper quadrant.
> ➤ Remember to *always* sketch the vectors in question. A small sketch of the vector can be very useful in determining components and angles of the vector.

9. TRUE or FALSE: A component of a vector is itself a vector.

10. Is the magnitude of a vector quantity a dimensionless number? What about its components? Why or why not?

Try It Yourself #8

Consider two vectors. $\vec{A}$ is 46.0 units long and makes an angle of 127° with respect to the $+x$ axis; $\vec{B} = 19.7\hat{\imath} + 23.8\hat{\jmath}$. Find the vector sum $\vec{C} = \vec{A} + \vec{B}$ of these two vectors. Write the answer in terms of magnitude and direction, as well as in component form.

Picture: To add these vectors, we need to break $\vec{A}$ into component form and then add it componentwise to $\vec{B}$.

Solve:

Draw a sketch. Include both vectors and the resultant vector.	
Find the x and y components of $\vec{A}$.	

Add the x component of $\vec{A}$ to the x component of $\vec{B}$, and do the same for the y components. This will give us the component form of the resultant.	$\vec{C} = -7.98\hat{\imath} + 60.5\hat{\jmath}$
Find the angle the resultant vector makes with respect to the x axis.	$\theta = 97.4$
Find the length of the resultant vector.	$C = 61.1$

Check: Since both vectors have positive y components, we know the result should be in the $+\hat{\jmath}$ direction, which it is. The resultant is also longer than either individual vector, but shorter than the sum of their lengths, as expected.

Taking It Further: If we were to calculate the vector $\vec{D} = \vec{A} - \vec{B}$, in what quadrant would the resulting vector point?

QUIZ

1. TRUE or FALSE: Two quantities having the same dimensions must be measured in the same units.

2. TRUE or FALSE: Two quantities having different dimensions can be multiplied, but they cannot be subtracted.

3. Why are all conversion factors dimensionless but not unitless?

4. Acceleration a has dimensions L/T^2, and those of velocity v are L/T. The velocity of an object that has accelerated uniformly through a distance d is either $v^2 = 2ad$ or $v^2 = (2ad)^2$. Which one is it? Why?

5. Consider two vectors: $\vec{A} = -23.0\hat{\imath} - 63.3\hat{\jmath}$ and $\vec{B}$ has a length of 43.1 and makes an angle of $-15°$ with respect to the positive x axis. What third vector $\vec{C}$ must be added to these two such that $\vec{A} + \vec{B} + \vec{C} = 0$? What are the components of $\vec{C}$? What angle does $\vec{C}$ make with respect to the x axis? What is the length of $\vec{C}$? What is the unit vector $\hat{C}$?

Chapter 2

Motion in One Dimension

2.1 Displacement, Velocity, and Speed

In a Nutshell

A **particle** is an idealized object whose motion can be described by the motion of a single point on an object. We can treat any object as a particle as long as we do not care about any effects related to its size, shape, or internal motion.

When some quantity is **constant**, that quantity does not change as *time* progresses. Rather, it maintains the same value.

Position is the location of a particle. **Displacement** is the change in a particle's position over some time interval.

In this chapter, we restrict ourselves to motion along a single straight line.

Average velocity is the total *displacement* of a particle divided by the time interval required for that displacement to occur. **Average speed** is the total *distance* traveled divided by the time interval required to travel that distance. In this graph, the average velocity of a particle as it moved from point P_1 to P_2 is shown.

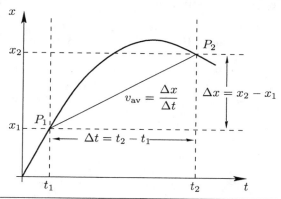

Instantaneous velocity is the limit of the average velocity as the time interval approaches zero. As points P_1, P_2, and P_3 move progressively closer to P, the slope of the average velocity line equals the tangent at P, which is the instantaneous velocity of the particle when it is at point P.

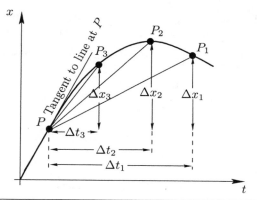

Physical Quantities and Their Units

Distance and displacement	meters, m
Time	seconds, s
Velocity and speed	m/s

Fundamental Equations

Displacement	$\Delta x = x_\mathrm{f} - x_\mathrm{i}$
Average speed	$\text{average speed} = \frac{\text{total distance}}{\text{total time interval}} = \frac{s}{\Delta t}$
Average velocity	$v_\mathrm{av} = \dfrac{\Delta x}{\Delta t}$
Instantaneous velocity	$v_x(t) = \lim\limits_{\Delta t \to 0} \dfrac{\Delta x}{\Delta t} = \dfrac{dx}{dt}$

Common Pitfalls

> ➤ The value of a quantity is distinctly different from a change or increment in that quantity. For example, 10:30 A.M. on Tuesday is a specific point (instant) in time, but 10.5 hours is a time interval.
> ➤ What makes things even more confusing is that we often start things at a time $t = 0$. In this case, $\Delta t = t_\mathrm{final} - t_\mathrm{initial} = t_\mathrm{final}$. Rather than always writing t_final, we just abbreviate and say that $\Delta t = t$. So you need to pay attention to the context to determine if the variable t is a particular instant in time, or a time interval that just happened to start at $t = 0$.
> ➤ Pay careful attention to signs. Signs can indicate direction in space. Velocity can be negative or positive.
> ➤ Speed is the magnitude of the (instantaneous) velocity, but average speed is not necessarily the magnitude of the average velocity. Average speed is defined as the total *distance* traveled divided by the time interval, while average velocity is defined as the *net displacement* divided by the time interval.

1. TRUE or FALSE: If a car is driven for 1 hour at an average velocity of +60 km/h, the distance it travels in that same 1-hour interval is 60 km.

2. Can the average speed of a particle be negative? Why or why not?

Try It Yourself #1

A car travels 40.0 km along a straight road at a speed of 86.0 km/h and then goes 40.0 km farther at a speed of 50.0 km/h. What is the car's average velocity for the entire trip?

Picture: This problem has two segments, each with a different speed. When calculating the total time or total displacement, we will need to add the time or displacement from each of the two segments.

Solve:

Write an *algebraic* expression for the car's average velocity.	
Find the total displacement Δx.	
Find the time required for the first leg of the trip. Units analysis can help you determine the form of the expression.	
Find the time required for the second leg of the trip.	
Determine the total time required for the trip by adding the previous two results.	
Substitute the calculated values of Δx and Δt to find the average velocity.	$v_{\text{av}} = 63.2 \text{ km/h}$

Check: We would expect the average velocity to be between the two given speeds, which it is. This average velocity is really a vector. The fact it is positive indicates that the displacement was in the $+x$ direction. If it were negative, then the displacement would be in the opposite direction.

Taking It Further: In this problem the average speed and the magnitude of the average velocity are the same. Explain why this is so.

Try It Yourself #2

Starting at time $t = 0$, a man walks east from his office to a restaurant, has lunch, and then walks west to his bank. His trip is shown on the graph of position versus time. (a) What was his average velocity from his office to the bank? (b) How much time did he spend at lunch? (c) How far is it from the restaurant to the bank? (d) At what point in the trip was he walking fastest?

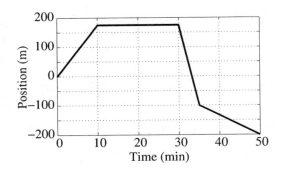

Picture: The average velocity between any two points is the slope of a straight line drawn between those two points. If the instantaneous slope at any point is zero, then the man must be momentarily at rest. Since the man starts off walking east, the positive positions must correspond to an easterly direction and the negative positions must correspond to a westerly direction. The man ended up at the bank.

Solve:

Find the average velocity of the man during his trip from the office to the bank. This will be total displacement divided by total time.	$v_{\text{av}} = -0.0667$ m/s
How long was the man eating lunch? He was not moving while he was eating.	20 min
What is the distance between the restaurant and the bank?	≈ 370 m
Estimate the speed of the man during each segment his trip.	
Use the results of the previous step to determine when he was walking fastest.	Immediately after leaving the restaurant the slope of the position vs. time graph is steepest. This is when he walked the fastest.

Check: The bank is much closer to the man's office than the total distance he walked would indicate.

Taking It Further: What is the man's average speed during his entire trip to the bank?

2.2   Acceleration

In a Nutshell

Average acceleration is the change in velocity divided by the time interval required for that change. **Instantaneous acceleration** is the limit of the average velocity as the time interval approaches zero.

It can often be helpful to draw motion diagrams when trying to analyze the motion of a particle. In a motion diagram, the position of a particle is drawn at equal time intervals, as if it were illuminated by a strobe light. At each position a vector is drawn whose length is proportional to the speed of the particle, and that points in the direction of the velocity. A second vector for the acceleration between two position points is drawn in the same manner.

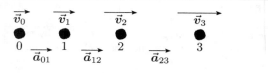

Physical Quantities and Their Units

Acceleration m/s^2

Acceleration due to gravity $g = 9.81 \text{ m/s}^2 = 32.2 \text{ ft/s}^2$

Fundamental Equations

Average acceleration $a_{\text{av},x} = \dfrac{\Delta v_x}{\Delta t}$

Instantaneous acceleration $a_x(t) = \lim\limits_{\Delta t \to 0} \dfrac{\Delta v_x}{\Delta t} = \dfrac{dv_x}{dt} = \dfrac{d^2 x}{dt^2}$

Common Pitfalls

> ➢ Pay careful attention to signs. Acceleration can be either positive or negative. When a particle is slowing down, the acceleration can be either positive or negative, depending on the choice for the positive direction. When a particle is slowing down, the acceleration and the velocity are always opposite in direction. In one-dimensional motion, opposite in direction goes hand in hand with opposite in sign.

> 3. TRUE or FALSE: The relation between velocity and position the same as the relation between acceleration and velocity.

4. Does a negative acceleration always mean an object is slowing down? Explain with at least one example.

Try It Yourself #3

A particle's position varies with time according to the expression $y = (4 \text{ m}) \sin(\pi t)$. Draw a motion diagram making sure to indicate the points where the position, velocity, and acceleration are zero and at their extreme values.

Picture: Velocity is the time derivative of position, and acceleration is the time derivative of velocity. To find maximum values, we take the derivative of the desired function with respect to time and set it equal to zero to find the time when the extrema occur.

Solve:

Find the velocity as a function of time by taking the derivative of the position with respect to time.	
Find the acceleration as a function of time by taking the derivative of the velocity with respect to time.	
By setting the velocity expression equal to zero, determine the first two times for extrema in the position.	$t = 0.5$ s, $t = 1.5$ s
By setting the acceleration equal to zero, determine the first two times for extrema in the velocity.	$t = 0$ s, $t = 1$ s

By setting the derivative of the acceleration equal to zero, determine the first two times for extrema in the acceleration.	$t = 0.5$ s, $t = 1.5$ s
Using the values above, draw your motion diagram. Make sure you get the directions of the velocity and acceleration vectors correct. Draw dots for the position at $t = 0, 0.25, 0.5, 0.75, 1, 1.25, 1.5, 1.75$, and 2.0 seconds.	

Check: Because the position varies sinusoidally, you should expect the position, velocity, and acceleration to repeat over time.

Taking It Further: Is the acceleration *always* proportional to the negative of the position? Why or why not?

2.3   Motion with Constant Acceleration

In a Nutshell

Motion with approximately constant acceleration commonly occurs in everyday life: the most common example is the motion of an object in free-fall near the surface of Earth. In this special circumstance, we can use the relations among acceleration, velocity, and displacement to derive a useful set of constant-acceleration kinematic equations. **Kinematic equations** describe the position, velocity, and acceleration of a particle as a function of time.

Important Derived Results

Constant acceleration equations

$$v_{av} = \tfrac{1}{2}(v_1 + v_2)$$
$$v_x = v_{0x} + a_x(\Delta t)$$
$$\Delta x = x - x_0 = v_{0x}(\Delta t) + \tfrac{1}{2}a_x(\Delta t)^2$$
$$v_x^2 = v_{0x}^2 + 2a_x(\Delta x)$$

Common Pitfalls

> ➢ The equations of motion with constant acceleration are used throughout the text. We recommend that you spend a few minutes now and commit them to memory. **Don't forget that they apply only to situations where the acceleration is constant.** These equations are developed with reference to the initial position x_0 and initial velocity v_0. In working a problem, it is often convenient to choose the origin at the initial position so that x_0 equals zero.

> ➢ Make sure you always label the origin and the positive direction of motion so you can maintain the distinction between positive and negative velocities and accelerations.

5. TRUE or FALSE: A ball is thrown straight up in the air. At its highest point, the acceleration of the ball is zero.

6. A bolt falls off its undercarriage when an elevator is (a) moving upward at constant speed, and (b) moving downward at the same constant speed. Assume the elevator is at the same height in both cases and disregard the effects of air resistance. Describe the motion of the bolt in both cases, including the speed with which each bolt hits the bottom of the shaft.

Try It Yourself #4

A model rocket starting from rest accelerates straight upward with a net acceleration three times that of gravity until it is 300.0 m above the ground. At that point its engine quits. How long after it was fired does the rocket hit the ground? Neglect air resistance.

Picture: This is a two-segment problem. The first segment has a constant acceleration equal to $3g$ upward, and the second segment will have an acceleration of g downward. The final velocity for the first segment will be equal to the initial velocity of the second segment. The position of the rocket at the beginning and end of each segment should be labeled. The total time the rocket is in the air will be the sum of the times required for each of the two segments.

Solve:

Sketch the situation, labeling the position, velocity, and acceleration at the start, on the way up, at the top, on the way down, and finally once the rocket hits the ground. Use subscripts to help you not confuse the position, velocity, and accelerations at the various points.	
Write *algebraic* constant-acceleration kinematic equations for the first segment. Use the variables identified in your sketch above.	
Write *algebraic* constant-acceleration kinematic equations for the second segment. Use the variables identified in your sketch above.	
Use the given acceleration and initial velocity to solve the equations in step two for the velocity when the rocket engine quits and the time when engine quits.	

Using the velocity when the engine quits, the height when the engine quits, and the given acceleration solve the equations in step three for the time interval required for the rocket to reach the ground from the time the engine quits.	
Find the total flight time by adding the time taken in each segment.	$\Delta t = 33.7$ s

Check: Because the upward acceleration is larger than the downward acceleration, and because the rocket is still traveling upward when the engine quits, the free-fall time should be larger than the engine burn time.

Taking It Further: How would you find the maximum height of the rocket?

Try It Yourself #5

Al sees Bob drive past his house at 20.0 m/s and realizes he must catch Bob to get the day's physics assignment. Al jumps in his car and starts it—this takes 15.0 s—and takes off after Bob. Suppose Al accelerates at a uniform rate of 1.40 m/s^2, and Bob continues driving at a constant speed of 20.0 m/s. Where and when does Al catch up to Bob?

Picture: This is a classic "chase" problem, but with two segments. We want to know both when and where Bob and Al have the same position. However, Al's motion has two segments: one in which he doesn't move and one in which he has a constant nonzero acceleration.

Solve:

Draw Bob and Al at their initial positions and again at their final positions. Include a coordinate axis and label the drawing with the kinematic parameters. Draw a separate diagram for each segment of the problem. Use subscripts to keep the variables describing Bob's and Al's motions separate.	
Write an *algebraic* equation of motion for Bob.	
Write an *algebraic* equation of motion for Al. Remember, for the first 15 seconds Bob moves away while Al doesn't move at all. This provides a relationship between the time variable for Bob's motion and the time variable for Al's motion.	
Set the positions of Al and Bob equal to solve for the time Bob drives before they meet.	$t = 54.4$ s
Now that you know the time, you can substitute this back into Bob's equation of motion to find where Al passes Bob.	$x = 1090$ m

Check: What is Al's speed when he catches Bob? It will have to be much larger than Bob's to make up for the fact that Al started later and from rest.

Taking It Further: What is required for Al to catch Bob such that Al has a final speed of 20.0 m/s, so they are both traveling the same speed when they first meet?

2.4   Integration

In a Nutshell

For situations in which the acceleration is not constant we cannot use the simple kinematics equations developed for constant acceleration. We must use the definitions of velocity and displacement and develop the more powerful integral expressions relating acceleration, velocity, and displacement.

Fundamental Equations

Integral form of the kinematic definitions

$$\Delta x = \int_{t_1}^{t_2} v_x \, dt$$

$$\Delta v_x = \int_{t_1}^{t_2} a_x \, dt$$

Common Pitfalls

> ➤ Remember the lower limit when evaluating integrals. The indefinite integral is so often equal to zero at the lower limit that it can be easy to forget. However, the initial conditions of any quantity are crucially important and should not be forgotten.

7. TRUE or FALSE: A particle with a position that is given by $x(t) = At + B$, with A and B properly dimensioned constants, is moving with a constant velocity.

8. You are analyzing a graph of the velocity of a particle versus time. Describe how to find the displacement and the acceleration of the particle at any given time.

Try It Yourself #6

The velocity of a particle is given by the equation $v = 3t^2 + 4$, where v is in meters per second when t is in seconds. Determine the displacement during the time interval from $t = 2.00$ s to $t = 5.00$ s.

Picture: The acceleration is not constant, so we cannot use the constant-acceleration equations. Instead, we use the general integral form for displacement.

Solve:

Write out the integral form for displacement in *algebraic* form.	
Use the given initial and final times and the expression for the velocity to solve for the total displacement.	$\Delta x = 129$ m

Check: Since the velocity is always positive for the given time interval, the displacement should also be positive.

Taking It Further: Why do we know the acceleration is not constant?

Try It Yourself #7

The acceleration of a particle is given by the equation $a = 4t - 5$, where a is in meters per second squared when t is in seconds. The velocity at $t = 0$ is -6.00 m/s. What is the velocity at $t = 4.00$ s?

Picture: The acceleration is not constant, so we cannot use the constant-acceleration equations. Instead, we use the general integral form for velocity.

Solve:

Write out the integral form for velocity *algebraically.*	

Substitute values and solve. Don't forget to include the initial velocity.	
	$v(4 \text{ s}) = 6.00 \text{ m/s}$

Check: The acceleration is increasing linearly with time, so the final velocity should be more positive than the initial velocity.

Taking It Further: How would you find the displacement during this same time interval?

QUIZ

1. TRUE or FALSE: If the average speed of a particle over a certain 3-s interval is 1.5 m/s, the average velocity over the same 3-s interval must be either +1.5 m/s or −1.5 m/s.

2. TRUE or FALSE: If at a certain instant the acceleration of a particle is positive, its velocity must also be positive.

3. At some instant a car's velocity is +15 m/s; 1 s later it is +11 m/s. If the car's acceleration is constant, what is its average velocity in this one-second interval? How far does the car go during that interval? Can you answer these questions if you don't know whether or not the acceleration is constant?

4. An elevator is moving upward at a constant speed when a bolt falls off its undercarriage. Does the bolt immediately descend, or does it continue to ascend until its velocity becomes zero? Why?

5. Starting from rest, a world-class sprinter can run 100 m in 10 s, but she cannot run 30 m in 3 s. Neither can she run 400 m in 40 s. Why not?

6. You're bicycling across central Oklahoma on a perfectly straight road. You started at 8:30 A.M. and you've covered 21.0 miles by 11:15 A.M., when your chain breaks. You haven't any spare parts with you, so you have to walk the bike back to the last town at a speed of 2.60 mi/h. If you get there at 1:30 P.M., what was (a) your displacement, (b) your average velocity, and (c) your average speed for the whole trip?

7. A man standing on a cliff 50.0 m above the level of the ground below throws a stone straight up in the air. The stone falls back past the edge of the cliff and strikes the ground 5.00 s after it was thrown. With what initial velocity did the man throw the stone?

Chapter 3

Motion in Two and Three Dimensions

3.1 Displacement, Velocity, and Acceleration

In a Nutshell

The concepts of position, displacement, velocity and acceleration easily extend to two and three dimensions. We now talk about the x, y and z components of each of these quantities.

The complete position, displacement, velocity, or acceleration vectors are the vector sum of their respective components.

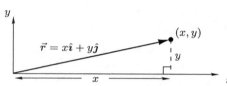

In many cases we can treat two- or three-dimensional motion as separate one-dimensional motion problems in the x, y, and z directions. The motion in one spatial dimension is linked to the motion in the other spatial dimensions by time, which must be the same in all spatial dimensions.

The concept of relative velocity is used to describe how one object moves with respect to two different frames of reference. If a person p sits in a train car C that is moving with a velocity of 10 m/s west to east relative to the ground G, then the person is also moving with that same velocity relative to the ground. If that person then gets up and walks to the front of the car at a speed of 1 m/s, and also to the right of the car at a speed of 1 m/s, then the person's velocity relative to the ground is 11 m/s west to east and 1 m/s north to south.

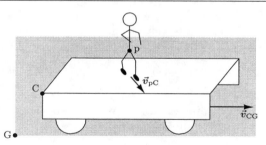

Algebraically, the relation between these velocities is given by

$$\vec{v}_{pG} = \vec{v}_{pC} + \vec{v}_{CG}$$

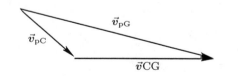

where $\vec{v}_{pG}$ is the velocity of the person relative to the earth, $\vec{v}_{pC}$ is the velocity of the person relative to the train car, and $\vec{v}_{CG}$ is the velocity of the train car relative to the earth. Note the order of the subscripts. As long as you maintain that order, you should be successful in your calculations.

Fundamental Equations

Position vector $\vec{r} = x\hat{\imath} + y\hat{\jmath}$

Displacement vector $\Delta\vec{r} = \vec{r}_2 - \vec{r}_1$

Average-velocity vector $\vec{v}_{\mathrm{av}} = \dfrac{\Delta\vec{r}}{\Delta t}$

Instantaneous-velocity vector $\vec{v} = \lim\limits_{\Delta t \to 0} \dfrac{\Delta\vec{r}}{\Delta t} = \dfrac{d\vec{r}}{dt}$

Relative velocity $\vec{v}_{\mathrm{pB}} = \vec{v}_{\mathrm{pA}} + \vec{v}_{\mathrm{AB}}$

Average-acceleration vector $\vec{a}_{\mathrm{av}} = \dfrac{\Delta\vec{v}}{\Delta t}$

Instantaneous-acceleration vector $\vec{a} = \lim\limits_{\Delta t \to 0} \dfrac{\Delta\vec{v}}{\Delta t} = \dfrac{d\vec{v}}{dt} = \dfrac{d^2\vec{r}}{dt^2}$

Common Pitfalls

> When you add vectors, remember that you cannot simply add their magnitudes. The equation $\vec{C} = \vec{A} + \vec{B}$ does not mean that $C = A + B$. Both of these equations are true only if $\vec{A}$ and $\vec{B}$ are in the exact same direction.

> The displacement between two points is defined as the straight-line magnitude and direction between them "as the crow flies." Its magnitude is the minimum distance actually traveled by the particle between the points. In many cases, the particle actually travels a greater overall distance than the actual displacement.

> It can be challenging to arrive at the equations for relative velocities. Make sure you pay careful attention to the order of the subscripts. If you do, you should be successful in your calculations.

1. TRUE or FALSE: When a particle is in motion, the difference between its position vector at time t_2 and its position vector at time t_1 is its displacement during the time interval $t_2 - t_1$.

2. Two displacement vectors $\vec{A}_1$ and $\vec{A}_2$ add to give a resultant of zero. What can we say about the two displacements?

Try It Yourself #1

At 3:00 P.M. you pass kilometer marker 160 as you are driving due south on Interstate 77. At kilometer marker 138, you turn off on U.S. 54. At 3:40 P.M. you have gone 14.0 km due southwest on U.S. 54. What was your average velocity (magnitude and direction) over this 40.0-min interval?

Picture: Where are you at the end of the 40 minutes? Your average velocity will be the total displacement divided by the time interval.

Solve:

Sketch the displacements during each segment of the problem. Make sure to include a coordinate system. It is often convenient for north and east to correspond to the $+y$ and $+x$ directions, respectively.	
Find the x and y components of each displacement vector. Let south to north be the $+x$ direction, and west to east be the $+y$ direction.	
Find the total displacement. First find the total x component of the displacement by adding the x components of each displacement vector calculated above. Then find the total y component of the displacement by adding the y components of each displacement vector calculated above.	
Find the direction of the displacement.	

Find the average velocity.	
	$\vec{v}_{av} = 50.1$ km/hr $17°$ west of south

Check: The magnitude of the average velocity is less than the average speed because the net displacement is less than the total distance traveled.

Taking It Further: If instead of following the roads, you were able to drive cross country "as the crow flies" direct to your ending point, but still taking the same amount of time, would your average velocity change? Why or why not?

Try It Yourself #2

The current at a certain point where the Wabash River is 80.0 m wide flows at 0.400 m/s. A swimmer sets out for a point directly across the river. (a) If the swimmer's maximum speed is 0.750 m/s relative to still water, in what direction should she swim to go directly to her goal? (b) How long will it take her to get there?

Picture: This is a relative-velocity problem. Once we find the swimmer's direction, we will use the component of her velocity perpendicular to the river to determine how long it will take her to cross the river.

Solve:

Draw a sketch to show the addition of veloci-ties. Call the $+x$ direction across the river and upstream the $+y$ direction.	

Use the relative-velocity vector relationship and your figure above to find the direction she should swim. The velocity of the swimmer with respect to the ground should be 100% across the river.	
From the direction of the swimmer and the speed of the swimmer relative to the water, find her across-stream speed. This is the x component of her velocity vector relative to the water.	$\theta = 32°$ upstream relative to straight across
Use the across-stream speed and the width of the river to determine how long it takes her to cross.	$\Delta t = 126$ s

Check: If the swimmer set off straight across the river, she would cross more quickly, because all her swimming velocity would be directed straight across. However, she would be much farther downriver by the time she reached the opposite bank.

Taking It Further: To achieve maximum speed with respect to the earth, in which direction should she head? Is this a reasonable way to cross the river? Why or why not?

3.2   Special Case 1: Projectile Motion

In a Nutshell

In projectile-motion problems, it is usually most convenient to let "up" be the $+y$ direction and $+x$ be the horizontal direction. Within this coordinate system, $a_x = 0$ and $a_y = -g = -9.81$ m/s^2. The horizontal and vertical equations of motion are synchronized by the time t.

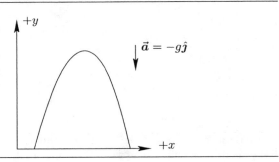

Important Derived Results

Horizontal displacement time dependence $\qquad$ $x(t) = x_0 + v_{0x}t$

Vertical displacement time dependence $\qquad$ $y(t) = y_0 + v_{0y}t - \frac{1}{2}gt^2$

$y(x)$ $\qquad$ $y(x) = (\tan\theta_0)\,x - \left(\dfrac{g}{2v_0^2\cos^2\theta_0}\right)x^2$

Range equation $\qquad$ $R = \dfrac{v_0^2}{g}\sin 2\theta_0$

Common Pitfalls

➤ The projectile-motion equations apply *only* when a particle is in free-fall, that is, when the only force affecting its motion is gravity. If air resistance affects its motion these equations do not apply.

➤ The projectile-motion equations can easily be modified to describe any two-dimensional motion with constant acceleration. Just remember to replace $-g$ with the appropriate acceleration in your situation.

➤ Consider a projectile thrown straight upward. Its velocity when it reaches the highest point in its trajectory is momentarily zero. But its velocity is changing direction, which means its acceleration is *not* zero. In fact, its acceleration is the same at the top as it is on the way up and on the way down: 9.81 m/s^2, directed downward.

➤ Consider a projectile thrown both upward and to the side, like a shotput. When such a projectile is at the top of its arc its velocity is *not* zero. Only the vertical component of its velocity is zero. The horizontal component of its velocity is constant throughout the motion.

➤ The range equation is valid *only* when the projectile's initial and final heights are identical.

3. TRUE or FALSE: The instantaneous-velocity vector is always in the direction of motion.

4. Suppose you have "sighted in" a rifle so that it will hit whatever the sights are aimed directly at on a target 150 m away over level ground. When shooting at a target 90 m away, should you aim above, below, or directly at the target? Why?

Try It Yourself #3

A ball rolls off a tabletop 0.900 m above the floor and lands on the floor 2.60 m away from a point that is directly under the edge of the table. At what speed did it roll off?

Picture: Remember the horizontal and vertical components of the ball's projectile motion can be treated separately. The vertical motion will provide the time the ball is in the air, which we can then use to find the initial horizontal velocity.

Solve:

Draw a sketch including the floor, the table, the trajectory of the ball, and your coordinate system. This should include the origin and the positive x and y directions.	
Write an *algebraic* expression for the motion of the ball in the y direction, and solve for the time when the ball hits the floor. The initial y velocity is zero as the ball rolls off the table edge. We always use the positive value for time if we have a quadratic equation to solve.	
Now that we know the time, we can use this in a similar expression for the motion of the ball in the x direction to solve for the initial x velocity. Since the initial y velocity was zero, the initial x velocity is the initial velocity.	6.07 m/s

Check: Although this initial velocity may seem a bit large, it is certainly physically possible to achieve without having to resort to rockets or other significant propulsion mechanisms.

Taking It Further: Let the table be slowly tipped downward at ever-increasing angles. Assuming the ball rolls off at that angle, can it still hit the same spot? What happens to the initial velocity?

Try It Yourself #4

In a game of American football, a quarterback throws a pass from his 20-yard line with an initial velocity of 22.0 m/s at an angle of 40° above the horizontal. The receiver starts running downfield from the 30-yard line 1.00 s before the pass is thrown. With what speed must he run in order to catch the pass? (Assume that the pass is caught at the same height above the ground at which it was thrown. If you need to convert units, 1 m = 1.09 yd.)

Picture: This is a classic "chase" problem, in which we are interested in the conditions required for two objects to be in the same place at the same time. For the receiver to catch the ball, the ball must return to its original height. We can use the projectile-motion equations to determine the horizontal displacement of the ball during its flight. We then equate the final position of the ball to the position of the receiver to solve for the receiver's speed.

Solve:

Draw a sketch showing the trajectory of the ball, and the receiver. Make sure to include the coordinate system you will use.	
Use the vertical equation of motion of the ball to find out how long it is in the air. Remember the initial and final heights of the ball are equal.	
Use the horizontal equation of motion of the ball to determine its final displacement from the quarterback.	

The receiver must cover 10 fewer yards than the horizontal displacement of the ball in one extra second. The displacement and time give us the speed the receiver must have.	
	$v = 11.1$ yd/s $= 10.2$ m/s

Check: This speed is comparable to the average speed of a world champion sprinter. That is why one of the key statistics for football players is the time to run a 40-yard dash.

Taking It Further: If the quarterback throws the ball at a greater angle above the horizontal at the same speed, will the receiver have to run faster or slower to catch the ball? Why?

3.3 Special Case 2: Circular Motion

In a Nutshell

An object moving in a circle at constant speed is experiencing **uniform circular motion**. The acceleration of an object in this case is always directed radially inward, toward the center point of the object's trajectory. This is **centripetal acceleration**. In the motion diagram, the position of an object is shown to move the same distance around the circular path in each time interval, so it is traveling at constant speed. You can see that $\vec{a}_{12} = (\vec{v}_2 - \vec{v}_1)/\Delta t = \Delta\vec{v}_{12}/\Delta t$ is directed toward the center of the circle. You should be able to convince yourself that the same is true for the other acceleration vectors as well.

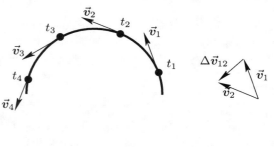

If the speed of the object is also changing, then in addition to the centripetal component of acceleration, there is also a tangential component to the acceleration. Consider the motion diagram of a simple pendulum. You can see that $\Delta\vec{v}_{45} = \vec{v}_5 - \vec{v}_4$ will result in an acceleration $\vec{a}_{45} = \Delta\vec{v}_{45}/\Delta t$ with some component toward the center of the motion. However, the acceleration is pointed down from where the center point would be because the object is slowing down as it rises to the top of its motion.

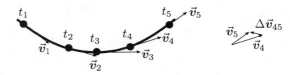

Important Derived Results

Centripetal acceleration $$a_\mathrm{c} = \frac{v^2}{r}$$

Tangential acceleration $$a_\mathrm{t} = \frac{dv}{dt}$$

Period of uniform circular motion $$T = \frac{2\pi r}{v}$$

Common Pitfalls

> For motion in one dimension, constant speed means that the acceleration is zero. This is *not* true for motion along a curved path. As long as the direction of the velocity vector is changing, even if the speed is constant, the acceleration vector is nonzero.

> Remember that if the speed of an object is changing while it is traveling along a curved path, then there is both a centripetal *and* a tangential component to the acceleration.

5. TRUE or FALSE: For a particle undergoing uniform circular motion, the instantaneous-acceleration vector equals zero.

6. TRUE or FALSE: For a particle in uniform circular motion, the average acceleration vector for a time interval is in the same direction as the change in the position vector for that time interval.

Try It Yourself #5

Exiting the freeway, you slow your car from a speed of 100 km/h to rest while traveling down a semicircular off ramp with a radius of curvature of 500 m. What is the magnitude of your acceleration as you exit the freeway and as you come to a stop at the end of the ramp, assuming your speed decreases at a uniform rate?

Picture: Because the car is slowing down *and* moving in circular path, there is both a centripetal and a tangential component to the acceleration. Because the centripetal acceleration depends on your speed, it will have different values for different parts of the off ramp.

Solve:

Sketch a motion diagram of the situation, showing the semicircular path and indicating the initial and final velocities and the tangential accelerations and centripetal accelerations.	

A constant tangential acceleration is responsible for the uniform change in speed of the car. Determine the total distance the car travels. As long as we consider only the tangential component of the acceleration, we can "straighten" the path of the car and use our one-dimensional kinematic expressions to determine the magnitude of the tangential acceleration.	$a_t = 0.246$ m/s^2
Find the centripetal acceleration. Because it depends on speed, we should find the centripetal acceleration at the start of the off ramp, and then at the end. For this you need both the initial and final speeds of your car.	$a_{c, \text{initial}} = 1.54$ m/s^2, $a_{c, \text{final}} = 0$ m/s^2
Find the magnitude of the total acceleration by adding the centripetal and tangential components in quadrature: take the square root of the sum of the squares of the individual acceleration components.	$a_{\text{tot,initial}} = 1.56$ m/s^2, $a_{\text{tot,final}} = 0.246$ m/s^2

Check: Even though the car slows at a constant rate, the initial acceleration is larger than the final acceleration.

Taking It Further: Would it be possible to maintain roughly the same acceleration magnitude during your entire trip on the off ramp? If so, how?

Try It Yourself #6

The hard disk drive in my computer rotates at 7200 RPM. If the disks are 10.0 cm in diameter, what is the centripetal acceleration of a point at the outside edge of the disk when the disk is at maximum speed?

Picture: To find the centripetal acceleration, you first must find the speed of the outside edge of the disk. You can find the circumference. You can also find the time for a single rotation from the given rotation speed.

Solve:

Determine the circumference of the hard disk.	
Convert the rotation rate of RPM into rotations per second.	
From the rotation rate, determine the time for a single revolution of the disk. Careful units analysis should help.	
From the circumference of the hard disk and the time for a single rotation, find the linear speed of a point on the outside edge of the disk.	
From the radius of the disk and the linear speed, determine the centripetal acceleration of a point on the edge of a disk.	$a_c = 28\,400$ m/s^2

Check: This is an amazingly high acceleration. When rotations are involved it is relatively easy to generate very high centripetal accelerations.

Taking It Further: Given that the human body can endure an acceleration equal to 7 to 9 times that of the acceleration due to the gravity of Earth before we pass out, what are the most dangerous times for stunt pilots?

QUIZ

1. TRUE or FALSE: The time it takes for a bullet fired horizontally to reach the ground is the same as if it were dropped from rest from the same height.

2. TRUE or FALSE: For a particle in uniform circular motion the average acceleration vector for a time interval is in the same direction as the change in velocity vector for that time interval.

3. Describe briefly the kind of motion a particle is undergoing when (a) the position vector changes in magnitude but not in direction; (b) the velocity vector changes in magnitude but not in direction; (c) the position vector changes in direction but not in magnitude; and (d) the velocity vector changes in direction but not in magnitude.

4. Explain how a particle moving in a circle with constant speed can have an acceleration.

5. Draw a motion diagram for a soccer ball that is kicked into the air from the ground, reaches a maximum height of 3.5 m, and on its way back down just enters the goal 30 m away at a height of 2.44 m. Include several positions and the x and y components of both the velocity and acceleration.

6. In a baseball game, a batter hits a fly ball directly toward the center fielder. The bat strikes the ball at a point 1.10 m directly above home plate, and the ball leaves the bat at a speed of 29.5 m/s at an angle of 35° above the horizontal. The center fielder is standing 116 m from home plate. If the center fielder starts running at the instant the ball is hit and if he catches the ball 0.5 m above the ground, how fast must he run to catch the ball? Assume that air resistance can be neglected.

7. A basketball player takes a shot when he is standing 24.0 ft from the 10.0-ft high basket. If he releases the ball at a point 6.00 ft from the floor and at an angle of 40° above the horizontal, with what speed must he throw the ball for it to hit the hoop?

Chapter 4

Newton's Laws

4.1 Newton's First Law: The Law of Inertia

In a Nutshell

Newton's first law: An object at rest stays at rest *unless* acted on by an external force. An object in motion continues to travel with constant speed in a straight line (with constant velocity) *unless* acted on by an external force.
When the motion of a particle is observed and measured, it must be observed and measured with respect to a specific reference frame. If no forces act on an object, any reference frame for which the acceleration of the object remains zero is an **inertial reference frame**.
Any reference frame moving with constant velocity with respect to an inertial reference frame is also an inertial reference frame.
A reference frame attached to the ground is not quite an inertial reference frame because of the small acceleration of the ground due to the rotation of Earth and Earth's orbit around the sun. However, these accelerations are on the order of 10^{-2} m/s^2 or less, so the surface of Earth is a reasonable approximation of an inertial frame of reference.

Common Pitfalls

> ➢ Remember that even if the speed of an object remains constant, it is possible for the object to be accelerating, which means that a net force must be acting on the object.

> ➢ Always choose a coordinate system that is fixed to an inertial reference frame. Newton's laws must be applied in inertial reference frames. For most of the problems considered in this chapter, the surface of Earth is considered an inertial reference frame and all coordinate systems will be fixed relative to Earth's surface.

1. TRUE or FALSE: If a particle moves with constant velocity, no forces are acting on it.

2. Why do you seem to be thrown outward when a car in which you are riding makes a sharp turn?

4.2   Force and Mass

In a Nutshell

A **force** is an external influence or action on an object that causes the object to change velocity, or accelerate relative to an inertial reference frame. A force is a vector quantity, with both a magnitude and a direction.

Forces are exerted on objects by other objects. **Contact forces**, such as friction or pushing a box with your hand, require direct physical contact between the two objects involved. **Action-at-a-distance** forces can act over large distances, without the need for physical contact. All four fundamental forces, described below, are action-at-a-distance forces.

The **gravitational force**—the very long-range interaction between particles due to their mass. It is believed by some that the gravitational interaction involves the exchange of hypothetical particles called gravitons.

The **electromagnetic force**—the long-range interaction between electrically charged particles involving the exchange of photons. This force is actually responsible for virtually all contact forces described in the text. We call them contact forces because the microscopic distances involved are usually extremely small compared to distances we typically see and experience in our macroscopic world. As a result, the objects *appear* to physically touch.

The **weak force**—the extremely short-range interaction between subnuclear particles involving the exchange or production of W and Z bosons. The electromagnetic and weak interactions are now viewed as a single unified interaction called the electroweak interaction.

The **strong force**—the short-range interaction between quarks and gluons that binds protons and neutrons together to form the atomic nuclei. Some physicists think that the strong and electroweak interactions are a single unified interaction.

Forces obey the **principle of superposition**. That is, if two or more separate forces act on an object, the result is the same as if a single, **net force**, equal to the vector sum of the separate forces, acted on that object. $\vec{F}_{net} = \vec{F}_1 + \vec{F}_2 + \cdots = \sum_{i=1}^{i=N} \vec{F}_i$.

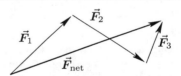

An object's mass is a measure of its **inertia**. The more massive an object, the greater its resistance to being accelerated.

Physical Quantities and Their Units

Unified atomic mass unit $1\ u = 1.660\,540 \times 10^{-27}$ kg

Common Pitfalls

> ➤ Remember that the net force is not an actual physical force. It is only the sum of the real forces acting on an object.

3. TRUE or FALSE: The forces that bind atoms together into molecules are electromagnetic in origin.

4. Explain how the frictional force between your hand and the carpet, as you rub your hand on the carpet, is electromagnetic in origin.

Try It Yourself #1

You and your friend are attempting to slide a box across a carpeted floor at constant velocity. If you push down on the box at an angle of 80° *below* the horizontal with a force of 20.0 N, the box has a weight of 50.0 N directed straight down, the floor pushes straight up on the box with a force of 17.0 N, and the force of friction is directed horizontally against the direction you push, with a magnitude of 24.9 N, with what force must your friend pull on the box so that the net force on the box is zero?

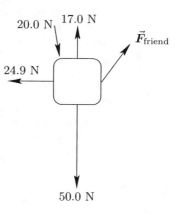

Picture: The sum of the forces must be zero because the box is not accelerating. Since force is a vector the individual forces must be added vectorally. The answer must have a magnitude and direction, or else be given in component form.

Solve:

Draw a sketch of all the force vectors described, using the head-to-tail method of vector addition. Include the force vector you are trying to find. Since the net force has to be zero, the head of the last vector must end up on the tail of the first vector.	
Find the horizontal (x component) and vertical (y component) of your pushing force on the box. Let "up" be the $+y$ direction and the horizontal direction you are pushing be the $+x$ direction.	
Add all the horizontal components of the vector forces together to find the x component of the pulling force of your friend. Because the frictional force is in the opposite direction of the motion of the box, it must be added with a negative sign. Remember that the sum of all the horizontal components has to be zero.	$F_{x,\text{friend}} = 21.4$ N

Add all the vertical components of the vector forces together to find the y component of the pulling force of your friend. Remember that the sum of all the vertical components has to be zero.	
	$F_{y,\text{friend}} = 52.7$ N

Check: If you accurately drew the force addition vectors in the first step, you should have found that $\vec{F}_{\text{friend}}$ needed to be up and to the right. Our numerical calculation agrees with that graphical representation.

Taking It Further: If you pushed with more force on the box, but in the same direction, would your friend have to pull with more or less force? Why? What else could you do that might guarantee your friend could pull with less force?

4.3   Newton's Second Law

In a Nutshell

Newton's second law: The acceleration of an object is directly proportional to the net force acting on it. The reciprocal of the mass of the object is the constant of proportionality.

$$\vec{a} = \frac{\vec{F}_{\text{net}}}{m}$$

The net force is the *cause*, and the acceleration is the *effect*.

A force of **1 newton** is defined as the force required to give a 1-kg mass an acceleration of 1 m/s^2.

Physical Quantities and Their Units

Force (SI units): newton	1 N $= 1$ kg $\cdot$ m/s^2
Force (U.S. system): pound	1 lb ≈ 4.45 N
Mass (U.S. system): slug	1 slug ≈ 14.6 kg

Fundamental Equations

Newton's second law
$$\vec{a} = \frac{\vec{F}_{net}}{m}$$

Common Pitfalls

> ➤ An acceleration does not create a force. If an object is accelerating it means that there must be a force from some external entity which is causing that acceleration.

5. TRUE or FALSE: The mass of an object may be determined in terms of the acceleration produced by a known force acting on it.

6. Suppose that only one force acts on an object. Can you tell in what direction the object is moving? Why or why not?

Try It Yourself #2

A string is used to spin a regulation hockey puck of mass 170 g in a circle of radius 2.50 m along an icy, frictionless surface at constant speed. If the tension in the string is 6.35 N, with what speed is the puck moving?

Picture: Newton's second law can be used to find the acceleration. Since the puck is moving in a circle, this is a centripetal acceleration, which can be used to find the speed of the puck.

Solve:

Use Newton's second law to find an *algebraic* expression for the acceleration of the puck. The *net* force acting on the puck is the tension in the string.	
Use the fact that the acceleration found above is a centripetal acceleration to determine the speed of the puck.	$v = 9.66$ m/s

Check: This speed is equivalent to just over 20 miles per hour, which may seem fast, but spinning objects can achieve reasonably high speeds quite easily.

Taking It Further: Identify the forces acting on the hockey puck. If you neglect friction, there should be three.

4.4   The Force Due to Gravity: Weight

In a Nutshell

The **weight** of an object is the gravitational force exerted upon it. On or near the surface of the moon, this is the gravitational force of the moon. On or near the surface of Earth, this is the gravitational force of Earth. We write the weight of an object as $\vec{F}_g = m\vec{g}$.

$\vec{g}$ is the **gravitational field** at the location of the mass m. In general, this field varies with position, although near the surface of Earth we will generally assume the gravitational field is uniform. We will typically consider the gravitational field created by the existence of large planetary-sized objects like Earth, the moon, and the sun.

Physical Quantities and Their Units

Gravitational field near the surface of Earth
$\vec{g} = (9.81 \text{ N/kg} = 9.81 \text{ m/s}^2 = 32.2 \text{ ft/s}^2)$
toward the center of Earth

Common Pitfalls

> ➢ Weight is not an intrinsic property of an object, but mass is. An object's mass does not change with location, but an object's weight does.

7. TRUE or FALSE: The weight of an astronaut in near-Earth orbit is zero.

8. Does your weight as measured in a stationary elevator change as the elevator accelerates upward or downward? Why or why not?

Try It Yourself #3

A net force of 7280 N acting on an object produces a net acceleration of magnitude 5.38 m/s^2 near the surface of Earth. What is the weight of the object?

Picture: We can use Newton's second law to find the mass, and use the mass to find the object's weight.

Solve:

Use Newton's second law to find the object's mass.	
Find the object's weight using the expression for weight.	$F_g = 13\,300$ N

Check: This mass and weight are roughly those for a small to mid-sized car.

Taking It Further: If this same force acted on the same object near the surface of the moon, would the object's acceleration remain the same? How about its weight? Explain.

4.5 Contact Forces: Solids, Springs, and Strings

In a Nutshell

When two solid objects touch, they exert forces on each other. We break those forces into two separate components. The **normal force** is the component of the force that is exerted perpendicular, or normal, to the interface between the two objects. The **frictional force** is the component of the force that is exerted in a direction parallel to the interface between the objects. Since these two forces are orthogonal, all forces between two objects can be treated as a combination of these two forces.

When a spring is stretched from its unstressed length by a distance x, the force it exerts is found experimentally to obey **Hooke's law**: $F_x = -kx$. The positive constant k is a measure of the stiffness of the spring. The negative sign indicates that if a spring is stretched, the force the spring exerts is to compress the spring back to its natural length, and if the spring is compressed, the spring exerts a force to stretch back out to its normal length.

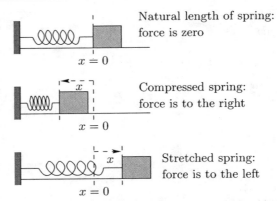

Natural length of spring: force is zero

Compressed spring: force is to the right

Stretched spring: force is to the left

Strings and ropes are used to pull things. Because they are flexible, they cannot push. We will generally assume that strings and ropes do not stretch. However, they can bend around other objects. The force that a string exerts on an object is equal to the **tension**, T, in the string.

Physical Quantities and Their Units

Spring constant k, with units of N/m

Fundamental Equations

Hooke's law $F = -kx$

Common Pitfalls

> The x in Hooke's law is the displacement from the *unstressed* length of the spring. So when doing Hooke's law problems, you should place the origin at the natural end of the spring.
> It is easy to get confused by the negative sign in Hooke's law. Simply remember that once compressed or stretched, the spring exerts a *restoring* force in the direction back toward its equilibrium length. Then examine your coordinate system to determine if this is a positive or a negative force.

9. TRUE or FALSE: According to Hooke's law, if a spring exerts a force of magnitude F when it is stretched a distance d, it will exert a force of magnitude $F/2$ when it is stretched a distance $2d$.

10. When a spring is compressed, it always exerts a positive force. Discuss this statement.

Try It Yourself #4

A 15.0-kg box sits on a spring as shown. A rope is attached to the ceiling and the top of the box. When the box is at rest, the tension in the rope is 75.0 N, and the spring is compressed a distance of 5.80 cm. What is the spring constant of the spring?

Picture: When at rest, the net force on the box is zero. There are a total of three forces acting on the box, which must sum to zero: gravity, the tension of the rope, and the spring.

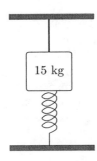

15 kg

Solve:

Depict the box as a dot. Then draw vectors for each of the forces acting on the box, and label each vector.	
Let "up" be the positive direction. Now apply Newton's second law, summing the forces to equal zero in this case. Rearrange this expression *algebraically* to solve for the spring constant k.	
Substitute values into your expression to find the spring constant.	
	$k = 1240$ N/m

Check: Your algebraic expression for k should demonstrate the following limiting properties: If the tension in the rope is equal to the weight of the the box, then the spring constant is zero. That is, the spring does not need to support the box. If the tension in the rope is zero, then the spring must support the full weight of the box.

Taking It Further: If the box is attached to the spring, under what conditions, if any, can the spring be stretched?

Try It Yourself #5

As is shown in the figure, a 25.0-kg traffic light is suspended from two light strands of wire with negligible mass. Determine the tension in each strand.

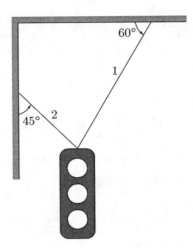

Picture: The traffic light is the object of interest, so we need to determine the forces acting on it. There are a total of three forces acting on the light, including its weight. Since the acceleration of the stoplight is zero, the vector sum of the forces should also be zero.

Solve:

Depict the traffic light as a dot. Then draw vectors for each of the forces acting on the light, and label each vector. The forces are the gravitational force and the tension from each of the wires. Remember that wires can only pull, and the force is in a direction parallel to the wires. Let the tension in wire 1 be $\vec{T}_1$, and let $\theta_1 = 60°$. Let the tension in wire 2 be $\vec{T}_2$, and let $\theta_2 = 45°$. Also draw a conventional coordinate system with $+x$ to the right and $+y$ up.	
Now you can apply Newton's second law, with the acceleration equal to zero. Since the forces act in more than one direction, you need to apply Newton's second law in component form, for both the x and y directions. Apply Newton's second law in the horizontal direction. This should allow you to relate the x component of $\vec{T}_1$ to the x component of $\vec{T}_2$ directly.	
Apply Newton's second law in the vertical direction to relate the y components of $\vec{T}_1$ and $\vec{T}_2$ to the weight of the light.	

Since we know the angles, we can write $T_{2,x} = -T_2 \cos \theta_2$, and $T_{2,y} = T_2 \sin \theta_2$, and similarly for $\vec{T_1}$. Use this fact to rewrite the equations of the previous two steps in terms of the magnitudes T_1 and T_2 directly, with the use of the known angles.	
Solve the two equations of the previous step for the two unknowns T_1 and T_2. Remember that you can easily find the weight using $F_g = mg$.	
	$T_1 = 180$ N, $T_2 = 127$ N

Check: Make sure you solved the system of equations correctly. The x components of the two tensions should have the same magnitude, and the y components of the tensions should add up to the weight of the light.

Taking It Further: Clearly the easiest way to hang a traffic light would be to use a single wire. Discuss the advantages of both the one- and two-wire systems.

4.6 Problem Solving: Free-Body Diagrams

In a Nutshell

One of the most critical problem-solving skills in physics is determining the forces acting on a single object or some system of objects. If two objects touch, they must exert a force on each other. In addition, there can be action-at-a distance forces that also need to be taken into account. The most common of these is an object's weight, due to the force of gravity, usually from Earth. Later in the course you will also have to consider the possibility of electric and magnetic forces. The following rules will help you determine what forces might be acting on an object.

Is the object in Earth's gravitational field?	Include the object's **weight** vector, $\vec{F}_g$ which always acts straight down, toward the center of Earth.
Is the object touching any other surfaces (table, wall, side of another block, etc.)?	For each surface the object is touching you should draw a **normal force** vector, $\vec{F}_n$ that is perpendicular to the surface. This force always tries to push the object away from the surface.
Do any of the surfaces above have friction?	For every surface that is not frictionless, you also need to draw a **frictional force** vector, $\vec{f}$, that is parallel to the surface, and in a direction that opposes the object's tendency to move. (You'll learn more about this in Chapter 5.)
Are there springs attached?	Determine if the spring is compressed or stretched from its equilibrium position. If the spring is compressed, then it pushes on the object. If the spring is stretched, then it pulls on the object.
Are there ropes or other pulling devices attached to the object?	You should have a **tension force** vector, $\vec{T}$, for each rope or pulling device. Remember that ropes always pull objects.
Is there some other applied force being delivered by a mechanism different from those above?	Include an **applied force** vector, $\vec{F}_{applied}$, for each such force.

A **free-body diagram (FBD)** is a picture that schematically shows all the forces acting on a system. Initially, we will treat our system as a single dot. A force vector representing each force acting on the object should start on the dot and point in the direction of the force. The length of the vector should be proportional to the magnitude of the force. Accelerations and velocities are *not* part of a FBD. If you feel the need to include these quantities (sometimes it can be very useful), then they should appear slightly separated from the actual FBD itself. A coordinate system should also be included.

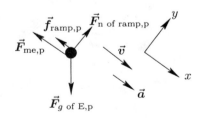

Every force should be labeled with at least one pair of subscripts. The first subscript of this pair should indicate what object is exerting the force. The second subscript of the pair should indicate the object that feels the force. In a free-body diagram, the second subscript of the pair should be the same for all forces.

Applying Newton's Second Law

Picture: Make sure you identify all the forces acting on the particle or system of interest. Then determine the direction of the acceleration vector of the particle, if possible. Knowing the direction of the acceleration vector will help you choose the best coordinate axes for solving the problem.

Solve:

1. Draw a neat diagram that includes the important features of the problem.
2. Isolate the object (particle) of interest, and identify each force that acts on it.
3. Draw a free-body diagram showing each of these forces.
4. Choose a suitable coordinate system. If the direction of the acceleration vector is known, choose a coordinate axis parallel to that direction. For objects sliding along a surface, choose one coordinate axis parallel to the surface and the other perpendicular to it.
5. Apply Newton's second law, $\vec{F} = m\vec{a}$, usually in component form.
6. Solve the resulting equations for the unknowns.

Check: Make sure your results have the correct units and seem plausible. Substituting extreme values into your symbolic solution is a good way to check your work for errors.

This method has been demonstrated, although not necessarily with this exact vocabulary, in the examples you have encountered so far in this chapter. You should re-examine those examples to find each of the above steps.

Common Pitfalls

> Except for action-at-a-distance forces, like gravity, two objects must be touching in order to exert a force on each other.

> Be sure to put only those forces on a free-body diagram for which you can identify an external physical source. Acceleration is not a force, and neither is the product of mass and acceleration ($m\vec{a}$). Any force on your free-body diagram should be identified as either an action-at-a-distance force or a contact force. With rare exceptions, the only action-at-a-distance force considered during the first semester of a general physics course is the gravitational force. In addition to action-at-a-distance forces, the diagram should contain at least one force vector for each thing that touches the object.

> Only those forces that act on an object go on that object's free-body diagram. Forces the object exerts on other things go on their free-body diagrams, not on its.

> The net force acting on an object is the name for the vector sum of all the forces that act on the object. It is not a separate force, and does not appear on a free-body diagram.

> The "centripetal" force does not belong on a free-body diagram. This is just a special name we give the physical force (maybe the tension in a string) that happens to be directed toward the center of a circular path.

> Be aware of the positive direction of the coordinate axes used in each problem. For example, the sign of the x component of a vector depends on the choice of the positive x direction (the direction of increasing x).

11. TRUE or FALSE: When all the forces that act on a particle and the mass of the particle are known, Newton's laws provide a complete description of the particle's motion.

12. A car is being driven up a long straight hill at constant speed. What forces act on it and what is the net force?

Try It Yourself #6

A 2.00-kg block on a horizontal, frictionless table is steadily pulled to the right by a string, as shown. In 5.00 s the speed of the block increases from zero to 10.0 m/s with constant acceleration. Determine the magnitude and direction of all forces acting on the block.

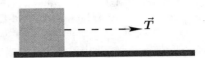

Picture: The string and table are both touching the block, so must exert forces on the block. We are not told otherwise, so we should also assume the block and table are on the surface of Earth, so we also need to include a weight. Because the block is accelerating, the net force will not be zero. The block will accelerate to the right, the direction it is being pulled. We will choose "right" as the $+x$ direction and "up" as the $+y$ direction, since all motion is horizontal.

Solve:

Draw a free-body diagram of the box. Make sure to include the coordinate axes.	
Determine the acceleration of the block. If it is sitting on the table, it will not accelerate vertically, but it can accelerate horizontally. Don't forget the sign of the acceleration. Use kinematic expressions to find a_x.	
Apply Newton's second law to the block. Write out the equation for Newton's second law in vector form for the block. Using your coordinate system and free-body diagram, write the equivalent equations for the x and y components separately, *in algebraic form*.	

Determine the weight of the box. Then solve the two equations of the previous step for $\vec{T}$ and the normal force of the table on the block.	
	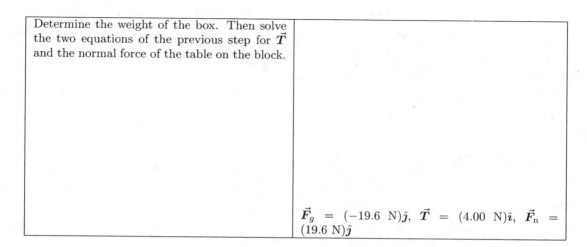 $\vec{F}_g = (-19.6 \ \text{N})\hat{\jmath}, \ \vec{T} = (4.00 \ \text{N})\hat{\imath}, \ \vec{F}_n = (19.6 \ \text{N})\hat{\jmath}$

Check: The table must push up on the block, and the string is clearly pulling to the right, so the direction of these forces is correct.

Taking It Further: If you pull on the string with the same force, but at an upward angle, what can you say about the resulting acceleration and normal force of the table on the block?

Try It Yourself #7

A 10.0-kg hanging ball is supported by a string that is attached to the ceiling. The object is pulled to the side by a second string connected to the object, as shown. Determine the tensions in the two strings.

Picture: The ball has three forces acting on it: the two tensions, and the ball's weight. The ball is stationary, so the net acceleration and force must be zero. Let the string attached to the ceiling be string 1, and the string attached to the wall be string 2.

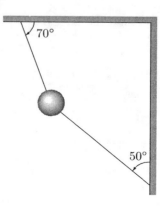

Solve:

Draw a free-body diagram of the ball, showing all three forces acting on the ball. Don't forget a coordinate system. Choose $+x$ to the right and $+y$ up.	
Apply Newton's second law for the ball. Write the equation for Newton's second law in vector form. Then write the equivalent component equations *in algebraic form* using the coordinate system chosen.	
Using the geometry of the problem, rewrite the components of $\vec{T}_1$ and $\vec{T}_2$ in terms of their magnitudes and the angles given.	
Rewrite the x- and y-component equations, using the expressions for the components of the tensions found in the previous step.	

Solve the two equations in the previous step for the two unknowns, T_1 and T_2.	
	$T_1 = 150$ N, $T_2 = 67.1$ N

Check: The x components of the two tensions cancel out, and the y components of the tension and the weight do add to zero.

Taking It Further: What angle for string 2 is required to minimize the tension T_2, and what is the minimum tension?

4.7 Newton's Third Law

In a Nutshell

Newton's third law: When two bodies interact, the force $\vec{F}_{BA}$ exerted by object B on object A is equal in magnitude and opposite in direction to the force $\vec{F}_{AB}$ exerted by object A on object B. $\vec{F}_{BA} = -\vec{F}_{AB}$

Common Pitfalls

> ➤ Notice that the action–reaction force pairs referred to by Newton's third law never act on the same object. Thus they never cancel when the net force acting on an object is calculated.

13. TRUE or FALSE: An object rests on a table top. The upward force by the table surface on the object and the downward gravitational force exerted by Earth on the object form an action–reaction pair and thus must be equal in magnitude and oppositely directed.

14. When you jump into the air, you have (for a short time) an upward acceleration. What external agent is exerting the upward force on you? What is the Newton's third law reaction force to the upward force exerted by this agent?

4.8 Problem Solving: Problems with Two or More Objects

In a Nutshell

A taut rope of negligible mass has the same tension throughout the rope. This is true even if the rope changes direction by passing over a frictionless surface.

If two or more objects are connected by a taut rope, then both objects must have the same tangential acceleration. This acceleration serves as the link between the two objects.

Applying Newton's Laws to Problems with Two or More Objects
Picture: Remember to draw a separate free-body diagram for each object. The unknowns can be obtained by solving simultaneous equations.
Solve:
1. Draw a separate free-body diagram for each object. Remember that if two objects touch, they exert equal but opposite forces on each other. The two ends of a rope also pull in opposite directions.
2. Apply Newton's second law to each object.
3. Solve the resultant equations, together with any equations describing interactions and constraints, for the unknown quantities.

Check: Make sure your answer is consistent with the free-body diagrams that you have created.

Common Pitfalls

> Two objects connected by a rope have the same tangential acceleration, but the absolute *direction* of that acceleration could be quite different.
> Remember that each connected object has its own distinct coordinate system. It is often very useful for these coordinate systems to be oriented at different angles.

15. TRUE or FALSE: If one object accelerating in the $+y$ direction is attached with a taut rope to a second object, the second object could be accelerating in the $-y$ direction.

16. Explain why the tension in a very light string connecting two objects is the same throughout its length.

Try It Yourself #8

Two blocks, connected by a light string as shown, are being pulled across a frictionless horizontal table top by a second light string. Block A has twice the mass of block B, the blocks are gaining speed as they move toward the right, and the strings remain taut at all times. Find the ratio T_1/T_2 of the two tensions.

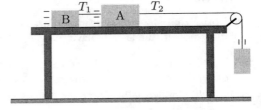

Picture: We do not know anything about the falling mass, so hopefully it will not appear in the final calculations. Because all three blocks are attached with light strings, they must have the same magnitude acceleration, but we don't know its value. We will need to draw separate free-body diagrams and apply Newton's second law to each block separately. To help keep things straight, it will be important to place a subscript A or B on each quantity that refers to block A or B, respectively.

Solve:

Draw a free-body diagram for each block. Block B should have three forces acting on it, and block A should have four forces acting on it. The string with tension T_1 exerts a force on both blocks, but in opposite directions. Each block also needs a coordinate system. In this case, it is convenient for each block to have the same coordinate system: $+x$ to the right, and $+y$ up.	
Write an *algebraic* expression for the acceleration in vector form for each block in terms of the forces on your free-body diagram. Use a subscript A for block A and a subscript B for block B.	
Apply Newton's second law to block A. Write an *algebraic* equation for Newton's second law in vector form for the block. Write the equivalent component equations using the coordinate system provided.	
Do the same thing for block B.	

Assuming the strings remain taut, but do not stretch, both blocks are constrained to accelerate together. This means we can set the magnitudes of their accelerations equal to each other.	
The magnitude of the tension in an ideal massless string is everywhere the same. This means you can set $T_{1,A} = T_{1,B}$.	
Using the equations for the x components of Newton's second law, we can now solve for the ratio of T_1/T_2. Note that we don't need to know the acceleration because it cancels out of the ratio.	$$\frac{T_1}{T_2} = \frac{1}{3}$$

Check: Since T_2 needs to accelerate both masses, and T_1 only needs to accelerate mass B, we expect that T_2 should be larger than T_1.

Taking It Further: Would this problem change if instead of having the hanging mass, the blocks were pulled by a hand? Why or why not?

Try It Yourself #9

Two blocks, connected by a light string as shown, are sliding toward the *left* across a frictionless horizontal table top. A second light string attached to block B pulls it to the right and as a result the blocks lose speed. Block A has two times the mass of block B and the strings remain taut at all times. Find the ratio T_1/T_2 of the two tensions.

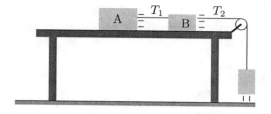

Picture: We do not know anything about the falling mass, so it should not appear in our final calculations. Because all three blocks are attached with light strings, they must have the same magnitude acceleration, but we don't know its value. We will need to draw separate free-body diagrams, and apply Newton's second law to each block separately. To help keep things straight, it will be important to place a subscript A or B on each quantity that refers to block A or B, respectively.

Solve:

Draw a free-body diagram for each block. Block A should have three forces acting on it, and block B should have four forces acting on it. The string with tension T_1 exerts a force on both blocks, but in opposite directions. Each block also needs a coordinate system. In this case, it is convenient for each block to have the same coordinate system: $+x$ to the right, and $+y$ up.	
Write an *algebraic* expression the acceleration in vector form for each block, in terms of the forces on your free-body diagram. Use a subscript A for block A and a subscript B for block B.	
Apply Newton's second law to block A *algebraically*. Write the equation for Newton's second law in vector form for the block. Write the equivalent component equations using the coordinate system provided.	

Do the same thing for block B.	
Assuming the strings remain taut, but do not stretch, both blocks are constrained to accelerate together. This means we can set the magnitudes of their accelerations equal to each other.	
The magnitude of the tension in an ideal massless string is everywhere the same. This means you can set $T_{1,A} = T_{1,B}$.	
Using the equations for the x components of Newton's second law, we can now solve for the ratio of T_1/T_2. Note that we don't need to know the acceleration because it cancels out of the ratio.	$$\frac{T_1}{T_2} = \frac{2}{3}$$

Check: Since T_2 needs to accelerate both masses, and T_1 needs only to accelerate mass B, we expect that T_2 should be larger than T_1.

Taking It Further: Why is this result different from that in the previous example?

QUIZ

1. TRUE or FALSE: When a large, fully loaded eighteen-wheeler truck runs head on into a small car, the magnitude of the force exerted by the truck on the car is the same as that exerted by the car on the truck.

2. TRUE or FALSE: If an object moves at constant speed, the net force acting on it must be zero.

3. Why do you seem to be thrown forward when a car in which you are riding stops abruptly?

4. Suppose a force $\vec{F}$ stretches a spring a distance Δx from its relaxed length, and another force $\vec{F}'$ stretches it by $2\,\Delta x$. How can you tell if $\vec{F}'$ is in fact equal to $2\vec{F}$?

5. When a pony pulls a stationary cart and sets it into motion, the force with which the pony pulls the cart forward exceeds the force with which the cart pulls back on the pony. Justify this statement or provide an argument why it is incorrect.

6. Two blocks are in contact on a frictionless, horizontal surface. The blocks are accelerated by a horizontal force $\vec{F}$ applied to one of them. Find the acceleration and the contact force for (a) general values of F, m_1, and m_2, and (b) for $F = 3.2$ N, $m_1 = 2$ kg, and $m_2 = 6$ kg.

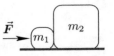

7. A small pebble of mass m rests on the block of mass m_2 in the apparatus shown. Find the force exerted by the pebble on m_2 if you let go of the masses and let them accelerate in Earth's gravitational field. $m_1 > m_2$.

Chapter 5

Additional Applications of Newton's Laws

5.1 Friction

In a Nutshell

Static friction f_s is present when two surfaces in contact do not slide on each other. Static frictional forces oppose any tendencies for the contacting surfaces to slide. For a given pair of contacting surfaces, the magnitude of the static frictional force ranges anywhere from zero to a maximum value that is proportional to the magnitude of the normal force: $f_s \leq \mu_s F_n$. As long as the box on the ramp in the figure does not start to slide down, there is a static friction force on the box directed up the ramp which prevents the box from sliding.

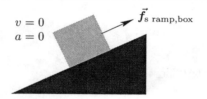

$v = 0$
$a = 0$

The force of **kinetic friction** f_k is present when skidding occurs; that is, when the surfaces in contact slide across each other, unless we are explicitly told that a surface is frictionless. The magnitude of the force of kinetic friction is given by the relation $f_k = \mu_k F_n$.

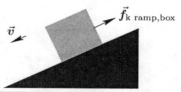

When an ideal, rigid wheel rolls at constant speed along an ideal, rigid horizontal road without slipping, no frictional force affects its motion. However, because real tires and roads continually deform and because the tread and road are continually peeled apart, the road exerts a force of **rolling friction**, $f_r = \mu_r f_n$, that opposes the motion.

The variables μ_s, μ_k, and μ_r are the coefficients of static, kinetic and rolling friction, respectively. For any pair of surfaces, the **coefficient of static friction** is *larger* than the **coefficient of kinetic friction**. In the figures above, notice that the arrow representing the force of static friction is larger than the arrow representing the force of kinetic friction. Values for μ_s and μ_k for a sampling of materials are listed in Table 5-1 on page 130 of the text. The **coefficient of rolling friction** is typically much smaller than the other two.

Solving Problems Involving Friction

Picture: Determine which types of friction are involved in solving a problem: static, kinetic, and/or rolling.

Solve:

1. Construct a free-body diagram with the y axis normal to (and the x axis parallel to) the contacting surfaces. The direction of the frictional force is such that it opposes slipping, or the tendency to slip.
2. Apply $\sum F_y = ma_y$ and solve for the normal force F_n. If the friction is kinetic or rolling, relate the frictional and normal forces using $f_k = \mu_k F_n$ or $f_r = \mu_r F_n$, respectively. If the friction is static, relate the frictional and normal forces using $f_s \leq \mu_s F_n$ (or $f_{s\,max} = \mu_s F_n$).
3. Apply $\sum F_x = ma_x$ to the object and solve for the desired quantity.

Check: Remember that coefficients of friction are dimensionless and that you must account for all forces (for example, tensions in ropes).

Physical Quantities and Their Units

Coefficient of friction μ coefficients of friction are unitless

Fundamental Equations

Static friction $f_s \leq \mu_s F_n; \; f_{s\,max} = \mu_s F_n$

Kinetic friction $f_k = \mu_k F_n$

Rolling friction $f_r = \mu_r F_n$

Common Pitfalls

➤ The value of the force of static friction f_s is *not* always equal to $\mu_s F_n$. The *maximum* value it can have is $\mu_s F_n$. Remember that while static friction is always directed so as to oppose relative motion between the contacting surfaces, static friction does not necessarily oppose motion. For example, when you start walking on a horizontal surface, like the floor of a classroom, it is the force of static friction exerted by the surface of the floor on the soles of your shoes that pushes you in the direction that you start to move. This force, which is in the forward direction, opposes relative motion between your foot and the floor by preventing your foot from sliding backward.

➤ When two surfaces come into contact, both a normal force component and a frictional force component of the contact force act on each surface. Either or both of these components can be zero in certain situations.

➤ You should *not* assume that the normal force on an object is equal to its weight. It is equal to the weight *only* for specific situations. The normal force is best determined by applying Newton's second law, $\vec{a} = \vec{F}_{net}/m$, in component form to the object. In doing this, choose one coordinate axis to be parallel to the normal force and the other parallel to the frictional force.

➤ For an object sliding along a flat surface the normal component of the acceleration is zero.

1. TRUE or FALSE: The magnitude of the static frictional force is directly proportional to the magnitude of the normal force as long as the static coefficient of friction remains constant.

2. When you are standing still and then start walking, what outside force acts upon you to cause your acceleration? Explain.

Try It Yourself #1

A block of wood of unknown mass m_1 rests on a 5.00-kg block, which in turn rests on a table top. The blocks are connected by a light string that passes over a frictionless peg. The coefficient of kinetic friction at both surfaces is $\mu_k = 0.330$. The upper block is pulled to the left by a 60.0-N force, and the lower block is pulled to the right by the string that passes around the peg. The blocks are moving at a constant speed. Determine the mass m_1 of the upper block.

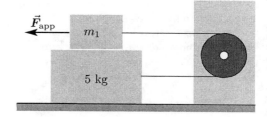

Picture: Because the blocks are sliding, this problem involves kinetic friction. The frictional force from the bottom block on the top block will be to the right, and the frictional force from the top block on the bottom block will be to the left, as they are third-law pairs. There is also a frictional force from the table on the bottom block. The tension on the string will exert the same magnitude force on each block. We will have to apply Newton's second law to both blocks, in both the horizontal and vertical directions, to solve the problem.

Solve:

Draw free-body diagrams of both blocks. The top block, which we will call block 1, should have five forces: the applied force, a normal force, a weight, a frictional force, and a tension. The bottom block, which we will call block 2, should have six forces: a normal and frictional force from the top block (third-law pairs of the forces the bottom block exerts on the top block), a normal and frictional force from the table, its weight, and the tension in the string. All your forces should have three subscripts. The first subscript should describe what kind of force it is: kinetic friction, normal, etc. The second subscript will be the object that exerts the force, and the third subscript should be the object that feels the force. Use these subscripts throughout the entire problem to avoid confusion. Use a coordinate system with the y axis vertical and the x axis to the right.	
Apply Newton's second law *algebraically* to the top block in *vector* form. What do we know about the acceleration of the top block?	

Use just the y components of Newton's second law, above, to find the normal force of the bottom block on the top block. There is no need to find a number here. An algebraic expression is just fine.	
Now write out the x component of Newton's second law applied to the top block. Make sure to relate the kinetic frictional force to the normal force you just solved for. Leave your expression in algebraic form, without numbers.	
Apply Newton's second law *algebraically* to the bottom block in *vector* form. What do we know about the acceleration of the bottom block?	
Work with the y component equation of Newton's second law for the bottom block to find the normal force of the table on the bottom block. Remember the normal forces that the blocks exert on each other are third-law pairs. Leave your result in algebraic form.	
Work with the x component equation of Newton's second law for the bottom block to solve for the tension in the rope. Remember to relate the normal force you just found to the frictional force between the table and the bottom block. Leave your result in algebraic form.	

Because we have a light string, the tension in the string is everywhere the same. In particular, the tension acting on the bottom block that we solved for in the previous step has the same magnitude as the tension acting on the top block. We also know that the forces of kinetic friction between the blocks are third-law pairs, and have the same magnitude. Use this fact to rewrite an *algebraic* expression for the tension T in the rope. This expression should only have masses, a coefficient of friction, and the acceleration due to gravity.	
Finally, substitute the expression you just found for tension into the x component expression for the top block. Rearrange the expression to solve for m_1 and substitute in the appropriate numbers	$m_1 = 4.51$ kg

Check: Calculate the value for the tension, too, and substitute all the numbers into the x-component equations for the top block and make sure the forces all add to zero.

Taking It Further: To accelerate the blocks, the applied force must be larger than 60.0 N. (a) If the blocks accelerate, will the frictional forces increase, decrease, or stay the same, and why? (b) If the applied force increases by 10.0 N to 70.0 N, will the tension in the string also increase by 10.0 N? (c) Why or why not?

Try It Yourself #2

A car is moving at 33.0 m/s down a hill that makes an angle of 20° with the horizontal. If the coefficient of static friction between the tires and the road is 0.580, what is the shortest distance in which the car can stop?

Picture: Draw both a sketch and free-body diagram of the car. The latter will include the weight of the car, as well as a normal force and frictional force. Use a rotated coordinate system so that the x axis is parallel to the incline of the hill and the y axis is perpendicular to the hill. This choice means the normal force is entirely in the $+y$ direction, and the acceleration will be only along the

x direction. Use Newton's second law to find the acceleration of the car, and kinematics to find the stopping distance. The shortest stopping distance will require the maximum static frictional force between the tires and the road.

Solve:

Draw a sketch of the situation. Include a vector indicating the direction the car is traveling, as well as a vector indicating the direction of the acceleration. Label the vectors.	
Draw a free-body diagram of the car, reducing the car to a particle represented by a single dot.	
Write Newton's second law in vector form. Also write the two separate equations, in *algebraic* form, for the x components and y components of the forces and accelerations.	
Consider the expression for the y components. The acceleration in the y direction is zero. Use the geometry of the problem to find the y component of the weight. Use all this information to solve for the normal force, F_{n}.	

Consider the expression for the x components. Use the geometry to find the x components of all the forces. Once you have those, solve for the acceleration in the x direction.	
Use kinematics to determine how far the car will travel until its velocity reaches zero. Since we know nothing about time, you will want to use the time-independent kinematic equation.	$\Delta x = 274$ m

Check: This is a reasonable distance, and the units are correct.

Taking It Further: If the driver of the car were to lock the wheels and go into a skid, how would the distance the car travels change, if at all, and why?

5.2 Drag Forces

In a Nutshell

When an object moves through a fluid such as air or water, the fluid exerts a **drag force** or retarding force that opposes the motion of the object. The drag force depends on the shape of the object, the properties of the fluid, and the speed of the object relative to the fluid. Unlike ordinary friction, the drag force increases as the speed of the object increases. At very low speeds, the drag force is approximately proportional to the speed of the object; at higher speeds, it is more nearly proportional to the square of the speed.

Fundamental Equations

Drag force $F_d = bv^n$

Important Derived Results

Terminal Speed in Free-fall $v_T = \left(\dfrac{mg}{b}\right)^{1/n}$

Common Pitfalls

> ➤ There are no rigid rules for determining the precise values of b and n for calculating drag forces. These values are determined empirically for each fluid and the precise geometry of the object traveling through the fluid. Any time you do a calculation without having done the rigorous measurements, all you have is a rough approximation.

3. TRUE or FALSE: The speed of an object falling under the influence of gravity through a fluid approaches a limiting value.

4. Can drag forces ever reverse your direction of travel? Why or why not?

Try It Yourself #3

A 5000-kg delivery truck rolling perfectly down a hill that makes a 35° angle with the horizontal reaches a terminal speed of 15.0 m/s. If the drag force is given by bv, what is the value for b?

Picture: Draw a picture of the problem, as well as a free-body diagram. Because the truck rolls perfectly, the coefficient of rolling friction will be zero. Since we have a terminal speed, the net force must be zero. We can use this to solve for the drag force, and then for the coefficient b.

Solve:

Draw a sketch and a free-body diagram of the situation. Choose a coordinate system in which the x direction is parallel to the hill and the y direction is normal to the hill.	

Write Newton's second law for the x components only. You will need to determine the x component of the weight of the car. Use this to find an expression for the magnitude of the drag force.	
Set the force calculated in the previous step equal to bv and solve for the drag coefficient.	$b = 1880$ N·s/m

Check: Your units for b should have dimensions of mass/time given this particular form for the drag force.

Taking It Further: If instead of a drag force, there was simply a standard frictional force acting on the truck, would the truck still have a terminal velocity? Why or why not?

5.3   Motion Along a Curved Path

In a Nutshell

The acceleration of a particle moving along a circular path has both a tangential component and a centripetal component. As a result, a particle moving along a curved path might experience a net force with both components. A centripetal component of the force directed toward the center of the circle is responsible for making the particle move in a circle, and the tangential component of the force will cause the particle to speed up or slow down.

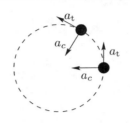

When cars travel around a banked curve, the normal force of the road on the car has a component directed in the centripetal direction, helping to keep the car moving in a circle.

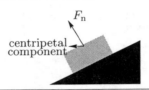

Solving Motion Along a Curved Path Problems

Picture: Remember that you should never label a force as a centripetal force on a free-body diagram. It should be labeled as a tension force, normal force, etc.

Solve:

1. Draw a free-body diagram of the object. Include coordinate axes at the point of interest along the path. One axis should point toward the center of the curved path (the centripetal direction), and the other will be tangential to the path, in the direction of motion.
2. Apply $\sum F_c = ma_c$ and $F_t = ma_t$, Newton's second law in component form.
3. Substitute $a_c = v^2/r$, and $a_t = dv/dt$, where v is the speed of the object.
4. If the object moves in a circle of radius r with constant speed v, use $v = 2\pi r/T$, where T is the time for one revolution.

Check: Make sure that your answers are in accordance with the fact that the direction of the centripetal acceleration is always toward the center of curvature and perpendicular to the direction of the velocity vector.

Common Pitfalls

> ➢ Do not treat the centripetal force as a separate force. It is the name given to the centripetal component of the net force. For motion along a general curved path, the centripetal direction is the direction toward the center of curvature of the path; thus for a circular path it is toward the center of the circle.

5. TRUE or FALSE: A spring force can have a centripetal component.

6. When a car travels around an unbanked curve, friction between the tires and the road provides the needed centripetal force. Is this friction kinetic or static? Explain.

Try It Yourself #4

A boy is swinging a 0.100-kg yo-yo in a vertical circle of radius 0.400 m by a light string, as shown. When the string is horizontal and the yo-yo is moving straight upward, the speed of the ball is 2.00 m/s. Determine the tension in the string and the rate of change of the yo-yo's speed at that instant.

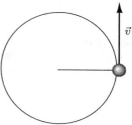

Picture: At the point of interest, the ball will have a centripetal acceleration directed to the left, toward the center of the circle. This acceleration can be used to find the tension in the string from Newton's second law applied to the horizontal direction. However, because the ball's weight acts tangential to the motion, in the downward direction, the ball will also have a changing speed, which can be found by applying Newton's second law to the vertical forces and acceleration.

Solve:

Draw a free-body diagram of the yo-yo. There should be only two forces.	
Write an *algebraic* expression of Newton's second law in vector form for the yo-yo.	
Apply Newton's second law in the horizontal (centripetal) direction to find the tension in the string. Because the ball is moving in a circle, we know this force causes a centripetal acceleration.	$T = 1.00$ N
Apply Newton's second law in the vertical (tangential) direction to find the rate at which the ball's speed is changing.	$a_{\mathrm{t}} = -9.81$ m/s^2

Check: At this point, gravity is the only downward force, so the tangential acceleration should simply be the acceleration due to gravity.

Taking It Further: Assume that the boy can swing the yo-yo in such a way that its speed remains constant. At what position of the yo-yo will the tension in the string be highest? Lowest? Explain.

Try It Yourself #5

Safety regulations require riders of roller coasters to wear seatbelts. However, it is possible to design a roller coaster that does a loop-the-loop, turning the riders upside down, in which the riders do not need to wear seatbelts. If the roller coaster has a radius of 20.0 m, what speed of the roller coaster is required at the top of the loop to keep the riders from falling out?

Picture: The sensation of falling occurs when your bottom loses contact with the seat of the roller coaster. That is, the seat no longer exerts a normal force on you. Therefore, the limiting case will be when the normal force of the seat on you is zero.

Solve:

Sketch the problem, and draw a free-body diagram of a roller-coaster rider at the top of the loop. Include both the normal force from the seat and the gravitational force.	
Apply Newton's second law in the vertical (centripetal) direction. This problem asks about the limiting case, in which the normal force from the seat is zero. Remember to use the centripetal acceleration to relate the speed of the roller coaster to the centripetal force. Use this to solve for the required speed of the coaster.	$v = 14.0$ m/s

Check: This is roughly half again as fast as the world's fastest sprinter, but easily achievable for amusement park rides.

Taking It Further: You solved for the threshold speed. Does the roller coaster need a speed above or below this threshold to keep its passengers from falling out? Why?

5.4 Numerical Integration: Euler's Method⋆

In a Nutshell

In some circumstances it is either very inconvenient or actually impossible to determine an object's position and velocity with an algebraic expression. In such situations we resort to numerical methods. Numerical methods require knowledge of the dependence of the acceleration on position, velocity, and time. The basic idea of **Euler's method** of numerical integration is to divide the time interval into a large number of short intervals and then to use the equations for motion with constant acceleration to solve for the changes in velocity and position of an object during the first short time interval. These new values of velocity and position are then used to update the value of the acceleration that is used to calculate the changes in the velocity and position during the second short time interval, and so forth.

To estimate the position x and velocity v at some time t, we first divide the interval from zero to t into a large number of small intervals, each of length Δt. The initial acceleration a_0 is then calculated from Newton's second law for the problem, using the initial position x_0 and velocity v_0. The position x_1 and velocity v_1 a time Δt later are estimated using the relations $x_{n+1} = x_n + v_n(\Delta t)$ and $v_{n+1} = v_n + a_n(\Delta t)$ with $n = 0$. The acceleration a_{n+1} is calculated using Newton's second law and the values for x_{n+1} and v_{n+1}, and the process is repeated. This continues until estimations for the position and velocity at time t are calculated.

Common Pitfalls

> ➢ When you use numerical methods to do integration, do not set the time intervals too small or the round-off errors will be too large. The challenge is to set them small enough to determine the value of the final velocity or position to the desired accuracy, but not so small that the round-off errors become excessive. One way to tell if you have set the time intervals small enough is to reduce them by a factor of 4 and then recalculate the results. If the results do not appreciably change, the size of the time interval is probably satisfactory.

⋆Optional material.

5.5   The Center of Mass

In a Nutshell

The motion of an extended, complex object or group of objects, like a person and bicycle jumping off a ramp, can be difficult to analyze. Wheels spin, the person may stand up on the seat, arms might wave in the air. However, one point, called the **center of mass** of the system, will travel through the air like the simple projectiles of Chapter 3, as if the mass of the entire system were located at the center of mass. Motion of the wheels, arms, etc. can then be described relative to the center of mass.

Although the individual particles of the system exert forces on one another, the acceleration of the center of mass is due only to the net *external* force. This is because the internal forces are third-law pairs, so the sum of the internal forces is always zero.

Solving Center-of-Mass Problems

Picture: Determining centers of mass often simplifies determinations of the motions of an object or system of objects. Drawing a sketch of the object or system of objects is useful when trying to determine a center of mass.

Solve:

1. Check the mass distribution for symmetry axes. If there are symmetry axes, the center of mass will be located on them. Use existing symmetry axes as coordinate axes where feasible.

2. Check to see if the mass distribution is composed of highly symmetric subsystems. If so, then calculate the centers of mass of the individual subsystems, and then calculate the center of mass of the system by treating each subsystem as a point particle at its center of mass.

3. If the system contains one or more point particles, place the origin at the location of a point particle. (If the ith particle is at the origin, then $\vec{r}_i = 0$.)

Check: Make sure your center-of-mass determinations make sense. In many cases, the center of mass of an object is located near the more massive and larger part of the object. The center of mass of a multi-object system or an object such as a hoop may not be located within or on any object.

Physical Quantities and Their Units

Linear mass density λ mass per unit length (kg/m)

Fundamental Equations

Center of mass, discrete particles

$$M\vec{r}_{\text{cm}} = m_1\vec{r}_1 + m_2\vec{r}_2 + \cdots = \sum_i m_i\vec{r}_i$$

Center of mass, continuous object

$$M\vec{r}_{\text{cm}} = \int \vec{r}\, dm$$

Mass of continuous, linear object

$$M = \int \lambda\, d\ell$$

Important Derived Results

Velocity of center of mass

$$M\vec{v}_{\text{cm}} = m_1\vec{v}_1 + m_2\vec{v}_2 + \cdots = \sum_i m_i\vec{v}_i$$

Acceleration of center of mass

$$M\vec{a}_{\text{cm}} = m_1\vec{a}_1 + m_2\vec{a}_2 + \cdots = \sum_i m_i\vec{a}_i$$

Common Pitfalls

> ➢ The center of mass of a system does not necessarily coincide with the position of any of the particles or objects that make up a system.

7. TRUE or FALSE: If the sum of the internal forces in a system remains zero, the acceleration of the center of mass necessarily remains zero.

8. How can the center of mass of an object be outside of the physical object itself? Give an example of such an object.

Try It Yourself #6

Find the center of mass of a thin rod of length L whose density increases linearly with the distance from one end according to the formula $\lambda = (2 \text{ kg/m}^2)x$, where x is the distance in meters from one end of the rod.

Picture: Use the integral form of the center-of-mass formula for the x direction: $Mx_{\text{cm}} = \int x\, dm$. Because the mass density is given to us, we do not need to calculate it.

Solve:

Make a sketch of the rod, with $x = 0$ at one end of the rod. Draw a small differential mass at some point along the rod, but not at either end or exactly in the middle. Assign the differential mass a differential length dx, and a coordinate x.	
Write an expression for the differential mass using the varying mass density provided and the differential length.	
Find the total mass by integrating the differential mass over the length of the rod.	

Now complete the center of mass calculation, first integrating to solve for Mx_{cm}, and then solving for x_{cm}.	
	$x_{cm} = \dfrac{2}{3}L$

Check: Because the object lies along a straight line, we expect the center of mass to be along the line. It should also be closer to the heavy end of the rod, which it is.

Taking It Further: Is it possible to have a non-uniform mass distribution in which the center of mass is located at the midpoint of the rod? How?

Try It Yourself #7

A uniform 0.300-kg meter stick, free to rotate about a peg located at its 20.0-cm mark, has a compact 0.200-kg weight fastened to the stick at the 80.0-cm mark as shown. As the meter stick rotates down to the vertical position with the 100-cm mark directly below the rotation axis, the stick–weight system's center of mass has a speed of 1.3 m/s. Determine the force exerted on the stick by the peg at this instant.

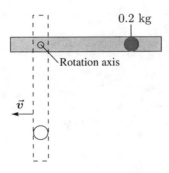

Picture: At the bottom of its motion the center of mass of the stick–weight system is experiencing circular motion and a centripetal acceleration. By drawing a free-body diagram of the stick–weight system and applying Newton's second law, the force of the peg on the stick can be found. Centripetal acceleration requires a radius, so we must find the center of mass of the stick–weight system, and use that to determine the radius of the motion for the center of mass. To find the center of mass of the stick–weight system, we can find the center of mass of the stick. Then assuming the mass of the stick is all at that location, we can find the center of mass of the stick–weight system using the standard summation expression.

Solve:

Find the center of mass of the uniform meter stick.	
Assuming the entire mass of the meter stick is located at its center of mass, determine the center of mass of the stick–weight system.	
Use the location of the center of mass, and the location of the rotation axis to determine the radius of the motion for the center of mass.	
Draw a free-body diagram of the stick–weight system when the stick is vertical. Represent the stick–weight system as a dot, located at the center of mass of the system.	
Apply Newton's second law to the stick–weight system, knowing that the center of mass experiences a centripetal acceleration. Use this to solve for the force of the peg on the stick.	
	$F = 6.91$ N

Check: This force is greater than the total gravitational force on the stick–weight system. It must be so in order to provide a net acceleration.

Taking It Further: If the weight moves closer to the peg, will the force of the peg increase or decrease, assuming that the speed of the center of mass remains constant? Why?

QUIZ

1. TRUE or FALSE: It is possible to survive if the center of mass of a hand grenade passes through the center of your head. (Not to be tried at home!)

2. TRUE or FALSE: It takes more force to keep a given pair of surfaces sliding against each other than it does to get them started.

3. Certain racing cars have an airfoil over the top of the car. The airfoil deflects air upward. What advantage is gained by deflecting the air upward when the car turns corners?

4. Why are curved roads banked?

5. Do the coefficients of either static or kinetic friction ever exceed the number 1? Why or why not?

6. A 3.00-kg particle is located at the origin, a 1.00-kg particle is located on the x axis at $x = 2.00$ m and a 2.00-kg particle is located in the xy plane at the point (2 m, 3 m) as shown. Determine the location of the center of mass of this three-particle system.

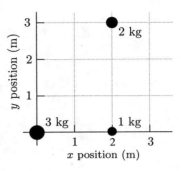

7. A 0.100-kg steel ball is suspended from the ceiling by a light string 0.400 m long. A physics instructor moves the ball to the side and then flicks it so that the ball passes through the lowest point in the arc with a speed of 0.800 m/s. Determine the tension in the string when the ball is just passing through the lowest point in its arc.

Chapter 6

Work and Kinetic Energy

6.1 Work Done by a Constant Force

In a Nutshell

In one dimension, the work done by a constant force on an object is the product of the force and the displacement of the point of application of the force. If the force and the displacement are in different directions as shown, then only the component of the force in the direction of the displacement is multiplied by the displacement. $W = F_x \Delta x = F \cos \theta \, \Delta x$.

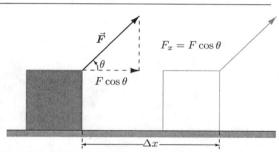

Work (which can be either positive or negative) is a scalar quantity and thus has no direction. When the force acts at right angles to the displacement, the work done by the force is zero.

When an object undergoes a displacement, each force acting on it may, in principle, do work. The **total work** W_{total} done on an object is the algebraic sum of the work done by each of the forces acting on it. An extended object can be modeled as a particle only if all its parts undergo identical displacements during any and all time intervals. That is, an object is a particle if it is perfectly rigid and moves without rotating.

The **work–kinetic-energy theorem**, which follows from Newton's second law, states that the total work W_{total} done on a particle equals its change in **kinetic energy** K.

Solving Problems Involving Work and Kinetic Energy

Picture: The way you choose the $+y$ direction or $+x$ direction can help you to easily solve a problem that involves work and kinetic energy.

Solve:

1. Draw the particle first at its initial position and second at its final position. For convenience, the object can be represented as a dot or a box. Label the initial and final positions of the object.

2. Put one or more coordinate axes on the drawing.

3. Draw arrows for the initial and final velocities and label them appropriately.

4. On the initial-position drawing of the particle, place a vector for each force acting on it. Accompany each vector with a suitable label.

5. Calculate the total work done on the particle by the forces and equate this total to the change in the particle's kinetic energy.

Check: Make sure you pay attention to negative signs during your calculations. For example, values for work done can be positive or negative, depending on the direction of the displacement relative to the direction of the force.

Physical Quantities and Their Units

Work   joules (J), dimensions of energy, or $\mathrm{ML}^2/\mathrm{T}^2$

Kinetic energy   joules (J), dimensions of energy, or $\mathrm{ML}^2/\mathrm{T}^2$

Fundamental Equations

Work by a constant force in one dimension $W = F_x \, \Delta x = F \cos\theta \, \Delta x$

Kinetic energy $K = \frac{1}{2}mv^2$

Work–kinetic-energy theorem $W_{\text{total}} = \Delta K = \frac{1}{2}mv_{\text{f}}^2 - \frac{1}{2}mv_{\text{i}}^2$

Common Pitfalls

> ➤ This chapter contains many words whose technical meanings differ from their meaning in everyday usage. These include **work, kinetic energy, potential energy,** and **power.** Be sure you understand the technical definitions.

> ➤ The work done by a force can be zero even if the force and the displacement are each nonzero. The work done by a force is zero when the displacement of the point of application of the force is either zero or is directed perpendicular to the force vector.

> ➤ Work is a scalar quantity, and thus it has no direction. The sign of a work term represents the sense of an energy change. When the work is positive, energy is transferred *to* the particle on which the force acts. When the work done on a particle is negative, energy is transferred *from* the particle.

> ➤ The work–kinetic-energy theorem can only be applied to a block sliding along a flat *frictionless* surface. If there is kinetic friction between the block and the surface then the block cannot be modeled as a particle. It follows that the theorem does not apply. Only if the surface is frictionless can the block be modeled as a particle.

1. TRUE or FALSE: Like kinetic energy, work is necessarily a positive quantity.

2. A block of mass m, released from rest, slides down a frictionless incline and reaches the bottom with speed v_{f}. What is the total work done on the block? How much work is done by each force acting on it? Explain.

Try It Yourself #1

Starting from rest, a 2.00-kg block is pulled up a frictionless 37.0° incline by a force $\vec{F}$ of 15.0 N directed up the incline, as shown. Determine the speed of the block after it has traveled a distance $d = 0.500$ m.

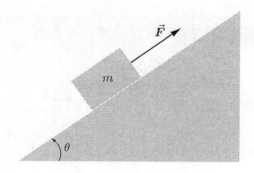

Picture: There are three forces acting on the block: the gravitational force, the normal force, and the applied force $\vec{F}$. By the work–kinetic-energy theorem the speed of the block is related to the total work done on the block by the sum of these forces.

Solve:

1. Draw two pictures of the particle on the ramp. The first one should show its initial position and the second should show its final position. Label them "initial" and "final" as appropriate. 2. Place a coordinate system on the drawing. Choose one with the $+x$ direction pointed up the ramp and the $+y$ direction perpendicular to the ramp. 3. Near each picture, draw a vector representing the velocity at the time shown. Label the vector. If you know the initial and/or final velocities, include that information on the pictures. 4. On the initial-position drawing of the particle, place a vector for each force acting on it. Accompany each vector with a suitable label.	
From your free-body diagram above, apply Newton's second law to find the normal force of the plane on the block.	
Find an *algebraic* expression for the work done on the box by the normal force. Remember that in this instance the normal force is perpendicular to the motion.	

Find an *algebraic* expression for the work done on the box by the applied force.	
Find an *algebraic* expression for the work done on the box by the gravitational force. Remember to include the angle between the force and the motion of the box.	
Use the work–kinetic-energy theorem to relate *algebraically* the change in speed of the box to the total work done on the box. Knowing that the initial speed is zero, solve for the final speed of the box.	$v_{\text{f}} = 1.26$ m/s

Check: Since the force of gravity effectively works against the applied force, the work done by gravity should be negative. Does your expression agree with that?

Taking It Further: Once the block travels a distance d, the applied force $\vec{F}$ is suddenly removed, so the block can slide back down the ramp. Once it slides back to its original position, will its speed return to zero? Why or why not?

Try It Yourself #2

A 5.00-kg block rests on a frictionless 53.0° incline, as shown. The block is pushed by a constant horizontal force $\vec{F}$ of 96.0 N. How far up the ramp does the block travel before its velocity is equal to 1.93 m/s?

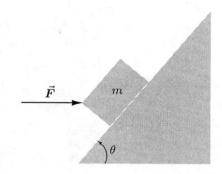

Picture: This problem, like the previous one, could be solved with the constant-acceleration kinematics. However, we will use the energy approach this time. There are three forces acting on the block: the gravitational force, the normal force, and the applied force $\vec{F}$. By the work–kinetic-energy theorem the speed of the block is related to the total work done on the block by the sum of these forces.

Solve:

1. Draw two pictures of the particle on the ramp. The first one should show its initial position and the second should show its final position. Label them "initial" and "final" as appropriate. 2. Place a coordinate system on the drawing. Choose one with the $+x$ direction pointed up the ramp and the $+y$ direction perpendicular to the ramp. 3. Near each picture, draw a vector representing the velocity at the time shown. Label the vector. If you know the initial and/or final velocities, include that information on the pictures. 4. On the initial-position drawing of the particle, place a vector for each force acting on it. Accompany each vector with a suitable label.	
Using your free-body diagram from the previous step, apply Newton's second law to find the normal force of the plane on the block.	
Find an expression for the work done on the box by the normal force. Remember that in this instance the normal force is perpendicular to the motion.	

Find an *algebraic* expression for the work done on the box by the applied force. Remember that the applied force and the motion of the box are not in the same direction.	
Find an *algebraic* expression for the work done on the box by the gravitational force. Remember to include the angle between the force and the motion of the box.	
Use the work–kinetic-energy theorem to relate *algebraically* the change in speed of the box to the total work done on the box. Knowing the initial and final speeds, the forces, and angles, solve for the distance up the ramp the box will travel.	$d = 0.500$ m

Check: Do you get the same answer if you apply Newton's second law and the constant-acceleration kinematic expressions? Which approach is easier?

Taking It Further: How can you increase the effectiveness of $\vec{F}$ in doing work on the box?

6.2 Work Done by a Variable Force—Straight-Line Motion

In a Nutshell

When a variable force does work, we can determine the work done by dividing the net displacement into a large number of very short displacements. We can estimate the work done in each short displacement by multiplying that displacement and the average value of the force during that displacement. The overall work done by a force acting over a displacement equals the sum of the work done by the force acting over a sequence of infinitesimal displacements that make up the displacement.

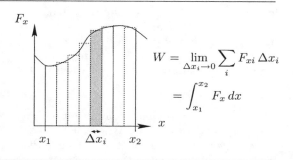

$$W = \lim_{\Delta x_i \to 0} \sum_i F_{xi} \, \Delta x_i$$

$$= \int_{x_1}^{x_2} F_x \, dx$$

Fundamental Equations

Work by a variable force—straight-line motion
$$W = \int_{x_1}^{x_2} F_x \, dx = \text{area under the } F_x\text{-versus-}x \text{ curve}$$

Work by a spring force
$$W_{\text{by spring}} = \tfrac{1}{2}kx_{\text{i}}^2 - \tfrac{1}{2}kx_{\text{f}}^2$$

Common Pitfalls

> The area "under" a force vs. position curve can be positive *or* negative. A positive displacement and a positive force, or a negative displacement and a negative force will result in *positive* work being done. However, a positive displacement and a negative force, or a negative displacement and a positive force will result in *negative* work being done on the particle.

3. TRUE or FALSE: Kinetic energy and work have the same dimensions.

4. A pendulum consists of a 1-inch-diameter steel ball of mass m suspended from the ceiling by a string. As the ball swings from its lowest position (where it has speed v_0) up to its highest position (where it reverses direction), what is the total work done on the ball by all the forces acting on it? How much work is done by each force acting on it? Explain.

Try It Yourself #3

A 2.00-kg particle is subjected to a single force F_x that varies with position as shown. From the graph, determine the work done by the force when the particle moves from $x = 0.00$ m to (a) $x = 4.00$ m and (b) $x = -3.00$ m. (c) If the particle is projected from the origin with an initial velocity of 1.00 m/s in the $+x$ direction, determine its speed when it is at $x = 4.00$ m.

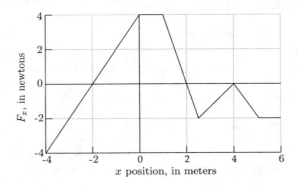

Picture: The work done by a force is equal to the area under the curve. When calculating the area, you must remember to include both the sign of the force and the displacement.

Solve:

To find the area, first divide the region between the curve and the x axis into rectangles and triangles. In the region from $x = -3.00$ m to $x = 4.00$ m, you will likely have five regions. You can do this on the figure above.	
Calculate the area of each region. Remember to include the sign of the force.	
To find the work done by the force for part (a), add only the area of the regions calculated above for $x = 0.00$ m to $x = 4.00$ m.	$W = 4.00$ J
To find the work done by the force for part (b), add only the area of the regions calculated above for $x = 0.00$ m to $x = -3.00$ m.	$W = -3.00$ J

To find the speed of the particle at $x = 4.00$ m use the work-kinetic energy theorem to relate the work done on the particle to the change in its kinetic energy. Solve the resulting expression for the final speed.	
	$v_f = 2.24$ m/s.

Check: Since positive work is done from the origin to $x = 4.00$ m, the particle's kinetic energy, and hence its speed, should increase, which they do.

Taking It Further: If the particle starts out with a velocity of 1.00 m/s in the *negative x* direction, determine its speed at $x = -4.00$ m by examining the graph. Explain.

Try It Yourself #4

A 5.00-kg block is attached to a spring with a spring constant of k. A constant applied force of $F = 30.0$ N pulls up on the block at a 14.0° angle and slides the block along a frictionless table as shown. Assuming the block starts with zero velocity when the spring is at its equilibrium position, and after being pulled a distance of 0.500 m it has a speed of 2.00 m/s, what is the spring constant of the spring?

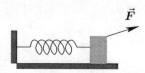

Picture: Four forces act on the block, but only two do any work. We will use the work–kinetic-energy theorem to find the spring constant of the spring.

Solve:

1. Draw two pictures of the box on the spring. The first one should show its initial position and the second should show its final position. Label them "initial" and "final" as appropriate. 2. Place a coordinate system on the drawing. Choose one with the $+x$ direction pointed to the right and the $+y$ direction directed upward. 3. Near each picture, draw a vector representing the velocity at the time shown. Label the vector with what we know about the initial final velocities. 4. On the initial-position drawing of the particle, place a vector for each force acting on it. Accompany each vector with a suitable label.	
Find an expression for the work done on the box by the normal force. Remember that in this instance the normal force is perpendicular to the motion.	
Find an expression for the work done on the box by the gravitational force. Remember that in this instance the gravitational force is perpendicular to the motion.	
Find an *algebraic* expression for the work done on the box by the applied force. Remember that the applied force and the motion of the box are not in the same direction.	
Find an *algebraic* expression for the work done on the box by the spring over the course of the entire displacement. This will require you to evaluate an integral.	

Use the work–kinetic-energy theorem to relate *algebraically* the change in speed of the box to the total work done on the box. Knowing the initial and final speeds, the forces, and the angles, solve for the spring constant of the spring.

$k = 36.4$ N/m

Check: Make sure your units are correct.

Taking It Further: Will there be a point in stretching the spring at which the speed of the box could be zero? If so, how would you find it?

6.3 The Scalar Product

In a Nutshell

The differential amount of work done by a force $\vec{F}$ acting over a small displacement $d\vec{\ell}$ is equal to $d\ell$ times the component of $\vec{F}$ in the direction of $d\vec{\ell}$: $dW = \vec{F} \cdot d\vec{\ell}$.

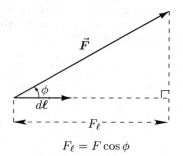

$$F_\ell = F \cos \phi$$

The **scalar** or **dot product** of two vectors is defined to be $\vec{A} \cdot \vec{B} = AB \cos \phi = A_x B_x + A_y B_y + A_z B_z$. The dot product of a unit vector and a vector $\vec{A}$ gives the component of $\vec{A}$ in the direction of the unit vector.

Power is the rate at which a force does work, or the work per unit time. In general, we use the term *power* to refer not just to the rate of doing work, but to any rate of energy transfer. The net power associated with the net force acting on a particle equals the *rate of change* of the particle's kinetic energy.

Physical Quantities and Their Units

Power	watt (W), with dimensions of Energy/Time
Power conversion factors	$1\ \text{W} = 1\ \text{J/s};\ 1\ \text{hp} = 550\ \text{ft} \cdot \text{lb/s} \approx 746\ \text{W}$
Energy conversion	$1\ \text{kW} \cdot \text{h} = 3.6\ \text{MJ}$

Fundamental Equations

Scalar Product	$\vec{A} \cdot \vec{B} = AB \cos \phi = A_x B_x + A_y B_y + A_z B_z$
Projecting $\vec{A}$ onto the x axis	$\vec{A} \cdot \hat{\imath} = A \cos \phi = A_x$
Definition of work	$W = \displaystyle\int_1^2 \vec{F} \cdot d\vec{\ell}$
Incremental work	$dW = \vec{F} \cdot d\vec{\ell}$
Power	$P = \dfrac{dW}{dt} = \vec{F} \cdot \vec{v}$
Kinetic energy and power	$P_{\text{net}} = \vec{F}_{\text{net}} \cdot \vec{v} = \dfrac{dK}{dt}$

Common Pitfalls

> ➤ A constant, non-zero net force can result in delivery of power that varies with time and position. Since a constant net force causes a constant acceleration, the velocity of the particle will vary continuously. As a result, the constant net force can result in the delivery of a varying amount of power.

5. TRUE or FALSE: The dot product of any two vectors is also a vector.

6. The tension in the string of a pendulum varies depending on the position and speed of the pendulum mass. Yet at every position and time, the tension delivers exactly zero power to the mass. How is this possible?

Try It Yourself #5

A particle undergoes a displacement from the coordinates $(3.00\ \text{m}, -1.00\ \text{m}, -4.00\ \text{m})$ to $(6.00\ \text{m}, -5.00\ \text{m}, -2.00\ \text{m})$. During this displacement a constant force of $\vec{F} = (-2.50\hat{\imath} - 3.75\hat{\jmath} - 1.50\hat{k})\text{N}$ acts on the particle. Find the work done on the particle by the force.

Picture: We will apply the definition of work, taking the three-dimensional dot product. Since the force is constant, there is no need to integrate and find differential amounts of work along the path.

Solve:

Make a sketch showing $\vec{F}$ and the initial and final positions. This can be tricky in three dimensions.	
Calculate the actual displacement vector $\vec{\ell}$.	
To find the work, calculate the dot product $\vec{F} \cdot \vec{\ell}$ using all three coordinates.	$W = 4.50$ J

Check: Note that having a negative displacement or a negative force along one component direction does not mean the work done by the force along that axis is negative. If both the displacement and the force are negative, the dot product, and hence the work, are positive.

Taking It Further: If the particle has a mass of m, can you find the final speed of the particle? How?

Try It Yourself #6

Find the power delivered by all forces present at the initial and final positions of **Try it Yourself** problems #2 and #4.

Picture: We already know the forces and velocities, so we simply need to apply the definition for power.

Solve:

Find the initial power of the normal force in problem #2.	$P_{\text{normal initial}} = 0$ W
Find the initial power of the gravitational force in problem #2.	$P_{\text{gravitational initial}} = 0$ W
Find the initial power of the applied force in problem #2.	$P_{\text{applied initial}} = 0$ W
Find the final power of the normal force in problem #2.	$P_{\text{normal final}} = 0$ W
Find the final power of the gravitational force in problem #2.	$P_{\text{gravitational final}} = -75.6$ W
Find the final power of the applied force in problem #2.	$P_{\text{applied final}} = 112$W
Find the initial power of the normal force in problem #4.	$P_{\text{normal initial}} = 0$ W
Find the initial power of the gravitational force in problem #4.	$P_{\text{gravitational initial}} = 0$ W

Find the initial power of the applied force in problem #4.	$P_{\text{applied initial}} = 0$ W
Find the initial power of the spring force in problem #4.	$P_{\text{spring initial}} = 0$ W
Find the final power of the normal force in problem #4.	$P_{\text{normal final}} = 0$ W
Find the final power of the gravitational force in problem #4.	$P_{\text{gravitational final}} = 0$ W
Find the final power of the applied force in problem #4.	$P_{\text{applied final}} = 58.2$ W
Find the final power of the spring force in problem #4.	$P_{\text{spring final}} = -36.4$ W

Check: The forces that oppose the motion actually remove energy from the boxes, so their powers are negative.

Taking It Further: Why is the power delivered by the normal force typically zero, but the power delivered by the gravitational force is sometimes nonzero? Could the power delivered by the gravitational force ever be positive? Why or why not?

6.4   Work–Kinetic-Energy Theorem—Curved Paths

In a Nutshell

The **work–kinetic-energy theorem** is readily extended to three dimensions. The total work done on a particle in three dimensions is equal to the change in the particle's total kinetic energy.

Fundamental Equations

Work–kinetic-energy theorem
$$W_{\text{net}} = \int_1^2 \vec{F}_{\text{net}} \cdot d\vec{\ell} = K_2 - K_1 = \Delta K$$

Common Pitfalls

> Remember that the work–kinetic-energy theorem only relates the forces to the initial and final kinetic energies, and hence speeds. It does not tell us anything about the time required.
> The kinetic energy only tells us about the speed of the object. We need more information to determine the object's *velocity*.

7. TRUE or FALSE: When several forces act on an *object*, the total work done by all of them is always equal to $\int_1^2 \vec{F}_{\text{net}} \cdot d\vec{\ell}$, where $\vec{F}_{\text{net}}$ is the net force acting on the object.

8. You pick a book up off your desk and place it on a bookshelf on the wall above the desk. What is the total work done by all forces acting on the book during this process? Explain.

Try It Yourself #7

A 2.00-kg particle undergoes a displacement from the coordinates (1.00 m, 3.00 m) to (5.00 m, 7.00 m). (a) How much work does the force $\vec{F} = 2x^2\hat{\imath}$ do on the particle? Assume that the force is in newtons when x is in meters. (b) If the particle starts with an initial velocity of $(4.00 \text{ m/s})\hat{\jmath}$, what is its final speed? (c) Determine the rate at which this force delivers energy to the particle at both its initial and final positions.

Picture: We need to integrate the force along the path to find the work done. Since the force is all in the x direction, we need only to integrate with respect to x because of the dot product. We will use the work–kinetic-energy theorem to find the final speed. Power will be $\vec{F} \cdot \vec{v}$.

Solve:

1. Draw a coordinate system showing the particle at its initial and final positions. Label them "initial" and "final" as appropriate. 2. Near each position of the particle, draw a vector representing the velocity at the time shown. Label the vector with what we know, if anything, about the initial and final velocities. 3. On the initial-position drawing of the particle, place a vector for the force acting on it.	
Use $W = \int_1^2 \vec{F} \cdot d\vec{\ell}$ to find the work done. We need only to integrate over the change in the x position because the force is only in the $\hat{\imath}$ direction.	$W = 82.7$ J
Use the work–kinetic-energy theorem to find the final speed of the particle.	$v_{\mathrm{f}} = 9.93$ m/s
Use the definition of power to find the power delivered to the particle by the force at the initial position.	$P_{\mathrm{initial}} = 0$
Knowing that the y component of the velocity remains constant, find the full vector velocity at the final position of the particle.	

Determine the value of the force at the final position.	
Use the definition of power to find the power delivered to the particle by the force at the final position.	$P_{final} = 455$ W

Check: This force is at least partially in the direction of motion along the entire path so we expect the kinetic energy, and hence speed, of the particle to increase.

Taking It Further: How would this problem change if there were also a force $F_y = 2y$ acting on the particle?

Try it Yourself #8

You and a friend are trying to determine the most efficient way to move a set of weights with total mass M from the floor to a shelf at a height of h and a horizontal distance of d from the weights. Your friend claims you will have to do more work if you move the weights straight up to a height of h first, and then over a distance d, all at constant speed, as compared to moving the weights along a direct diagonal path. Is he right?

Picture: Since the weights start and end with zero speed, but certainly move from one point to another, a brief acceleration at each end of the path is required. We often neglect these small, hard-to-quantify regions. Furthermore, in this case if the magnitude of the acceleration at the beginning and end of the weight's motion is the same, the net work required for these two accelerations is zero. Constant speed implies that the net work on the weights has to be zero. The only two forces acting on the weights during their trip are gravity and you. To find the work done on the weights, we will use the definition $W = \int_1^2 \vec{F} \cdot d\vec{\ell}$ for each path.

Solve:

1. Draw two pictures of the weights, one for each path the weights can take. Each picture should show their initial and final positions, as well as the path taken to move the weights. 2. Place a coordinate system on the drawing. Choose one with the $+x$ direction pointed horizontally from the weights to the shelf and the $+y$ direction up. 3. Near each picture, draw a vector representing the velocity at the time shown. Label the vector. If you know the initial and/or final velocities, include that information on the pictures. 4. On the initial-position drawing of the weights, place a vector for each force acting on it (there should be two). Accompany each vector with a suitable label.	
Find the work done by the gravitational force as the weights are lifted straight up.	
Find the work done by the gravitational force as the weights are moved horizontally over to the shelf.	
Find the total work done by the gravitational force during this "long" trip.	
To move the weights at constant speed (no net work), determine the total work you must do on the weights during this "long" trip.	$W = Mgh$
Find the total work done by the gravitational force as the weights are moved diagonally to the shelf.	

To move the weights at constant speed, determine the total work you must do on the weights during this diagonal trip. The work you must do on the weights is the same, regardless of the path taken.	
	$W = Mgh$

Check: Since the total displacement is the same in both cases, and since the only other force acting is the same in both cases, the total work done by that other force will be the same in both cases.

Taking It Further: What if you first lifted the weights to a height of $2h$ before resting them on the shelf? Would the work you had to do on the weights change? If so, how? We all know from experience that your friend is right, and it takes more "work" to move the weights along the "long" path than the diagonal. Reconcile this common knowledge with our calculations here.

6.5   Center-of-Mass Work⋆

In a Nutshell

When an object cannot be modeled as a particle, we can still apply the work–kinetic-energy theorem. However, it needs to be applied to the motion of the *center of mass* of the object. That is, we consider the displacement of the center of mass of the object. Furthermore, in this section we also consider only the **translational kinetic energy** of the object—the kinetic energy related to the speed of the center of mass.

Fundamental Equations

Center-of-mass work–translational-kinetic-energy relation $\displaystyle \int_1^2 \vec{F}_{\text{net ext}} \cdot d\vec{\ell}_{\text{cm}} = \Delta K_{\text{trans}}$

⋆Optional material.

QUIZ

1. TRUE or FALSE: An athlete throws a ball. To throw the ball again so that it moves with twice the speed requires that the athlete do twice as much work on the ball.

2. TRUE or FALSE: When several forces act on a *particle*, the total work done by all of them is always equal to $\int_1^2 \vec{F}_{\text{net}} \cdot d\vec{\ell}$, where $\vec{F}_{\text{net}}$ is the net force acting on the particle.

3. A car of mass m travels up a long, straight hill of height h at constant speed. Assume air drag is negligible. How much work is done by each force acting on the car? What is the total work done on it by external forces? Explain.

4. When you get up from your chair and start walking toward the door, your kinetic energy increases. What force does work on you to cause this increase in your kinetic energy? Where does this kinetic energy come from?

5. What must be true in order for a force to do negative work on a particle?

6. A 2.00-kg particle is subjected to a single force F_x that varies with position as shown. From the graph, determine the work done by the force when the particle moves from $x = 0.00$ m to (a) $x = 6.00$ m and (b) $x = -4.00$ m.
(c) If the particle is projected from $x = -4.00$ m with an initial velocity of 5.00 m/s in the $+x$ direction, determine its speed when it is at $x = 6.00$ m.

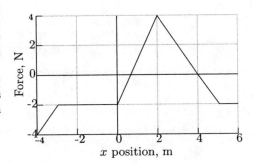

7. In Austria, there once was a 5.6-km-long ski lift. It took about 60 minutes for a gondola to travel up its length. If there were 12 gondolas going up, each with a cargo of mass 550 kg, and if there were 12 empty gondolas going down, and the angle of ascent was 30°, estimate the power P required of the engine to operate the ski lift.

Chapter 7

Conservation of Energy

7.1 Potential Energy

In a Nutshell

The work done by a **conservative force** on a particle is independent of the path taken as the particle moves from one point to another. It follows that a second definition of a **conservative force** could be: A force is **conservative** if the work it does on a particle is zero when the particle moves around *any* closed path, returning to its initial position. The work along path 1 plus the work along path 2 should add to zero for a **conservative** force. The force of gravity, and the spring force are examples of conservative forces. However, many forces like friction, pulling on a rope, etc. are **not** conservative.

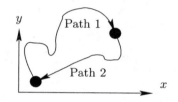

The work done by an external agent on a system of particles can result in a change of kinetic energy of the system, be stored as *potential energy* within the system, or result in changes of thermal, chemical or other internal energies which we will discuss later in this chapter. **Potential energy** is a form of energy associated with the relative *positions* of different parts of a system. This contrasts with kinetic energy, which is associated with *motion*.

The work done by a conservative force is used to define a **potential energy function U** associated with that force. The **change in potential energy** associated with a conservative force is the negative of the work done by that force.

Only *changes* in potential energy are important. As a result, we are generally free to decide what system configuration corresponds to zero potential energy. However, there are often preferred configurations for the zero of the potential energy. One example of this is the potential energy due to a spring. When we define $U = \frac{1}{2}kx^2$, we also implicitly define $U = 0$ to correspond to the relaxed state of the spring, which we define to be at $x = 0$.

The **gravitational potential energy** of a system of particles in a uniform gravitational field is the same as if the entire mass of the system were concentrated at the system's center of mass. As such, changes in the gravitational potential energy of the system of particles are associated only with changes in the position of the system's center of mass.

Physical Quantities and Their Units

Potential energy U units of joules (J), dimensions of energy

Fundamental Equations

Change in potential energy $\Delta U = U_\mathrm{f} - U_\mathrm{i} = -W = -\displaystyle\int_1^2 \vec{F} \cdot d\vec{\ell}$, where $\vec{F}$ is conservative

Important Derived Results

Potential energy function near Earth's surface $U_g = U_0 + mgy$

Potential energy of a spring $U_{spring} = \frac{1}{2}kx^2$

Common Pitfalls

> ➤ Only changes in potential energy have physical meaning. The choice of the reference point for potential energy is always arbitrary, so choose a reference point that is convenient for a particular situation.
> ➤ Although kinetic energy is always positive, potential energy can be either positive or negative.
> ➤ Do not associate potential energy with a single particle. Potential energy depends on the system configuration and is associated with the entire system. If only one particle in a system moves, it is common parlance to say that the particle's potential energy has changed, but in reality it's the potential energy of a larger system that has changed.
> ➤ Be aware of the sign in the definition of potential energy. When the displacement is in the direction of a conservative force, the change in potential energy is *negative*. For example, when an object moves downward, in the direction of the gravitational force, the gravitational potential energy decreases.

1. TRUE or FALSE: Particles that attract via conservative forces have more potential energy when they are close together than when they are far apart.

2. Explain why a spring stretched a distance d has the same potential energy as a spring compressed the same distance d, and that the potential energy is positive in both cases.

Try It Yourself #1

Find the total work done by the force $\vec{F} = 4xy\hat{i} - 3x^2\hat{j}$, where $\vec{F}$ is in newtons when x and y are in meters, around the closed rectangle defined by the diagonal vertices at $(3 \text{ m}, 2 \text{ m})$ and $(5 \text{ m}, -4 \text{ m})$. Is this a conservative force?

Picture: Work is $\int_1^2 \vec{F} \cdot d\vec{\ell}$. We will treat each segment of the rectangle separately. If the total work is zero, then the force might be conservative.

Solve:

Draw the rectangle described, including a co-ordinate system and the coordinates of all the vertices.	

Determine the work done by the force on a particle as it travels from $(3\text{ m}, -4\text{ m})$ to $(3\text{ m}, 2\text{ m})$. Your $d\vec{\ell}$ will be $dy\hat{\jmath}$, and as you travel along this segment, the value of x has a constant, nonzero value.	
Determine the work done by the force on a particle as it travels from $(3\text{ m}, 2\text{ m})$ to $(5\text{ m}, 2\text{ m})$. This time your $d\vec{\ell}$ will be $dx\hat{\imath}$, and as you travel along this segment, the value of y has a constant, nonzero value.	
Determine the work done by the force on a particle as it travels from $(5\text{ m}, 2\text{ m})$ to $(5\text{ m}, -4\text{ m})$. Make sure, as you evaluate this integral, that you have a *negative* displacement parallel to the y axis. The value of x has a constant, nonzero value.	
Determine the work done by the force on a particle as it travels from $(5\text{ m}, -4\text{ m})$ to $(3\text{ m}, -4\text{ m})$. Make sure, as you evaluate this integral, that you have a *negative* displacement parallel to the x axis. The value of y has a constant, nonzero value.	
Find the total work done along the path by adding the results of the previous 4 calculations. Since this work done during the round trip is nonzero, we can say that this is definitely *not* a conservative force.	$W_{\text{round trip}} = 480\text{ J}$

Check: 1 N·m= 1 J, so the units work out. If you traverse the rectangle in the opposite direction you should get −480 J.

Taking It Further: Explain why the work is negative as we travel up the left side of the rectangle, and positive as we travel down the right side, but the work is positive for *both* the top and bottom segments, even though we traverse them in opposite directions.

Try It Yourself #2

A box of mass m rests on a frictionless table. Initially, it compresses a massless spring with a constant k a distance d. The spring is then let go. How fast is the mass moving when the spring reaches its equilibrium length?

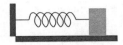

Picture: Treat the box as a particle. The spring will lose potential energy. The change in potential energy equals the negative work done by the conservative spring force. This work done will equal the change in kinetic energy of the box.

Solve:

1. Draw two pictures of the particle attached to the spring. The first one should show its initial position and the second should show its final position. Label them "initial" and "final" as appropriate. 2. Place a coordinate system on the drawing. Choose one with the $+x$ direction pointed to the right, and the y direction up. 3. Near each picture, draw a vector representing the velocity at the time shown. Label the vector. If you know the initial and/or final velocities, include that information on the pictures. 4. On the initial-position drawing of the particle, place a vector for each force acting on it. Accompany each vector with a suitable label.	
Determine the change in potential energy of the spring.	
Find the work done by the spring on the particle by relating the work to the change in potential energy.	
Equate the work done by the spring to the change in kinetic energy. Use this relationship to solve for the final speed of the particle.	$v_\mathrm{f} = \sqrt{\dfrac{k}{m}}\,d$

Check: Convince yourself the units of this answer will be in m/s.

Taking It Further: If the mass is attached to the spring, how fast will it be moving after the spring continues to move and ends up stretched a distance d? Will this answer be different if the particle is not attached to the spring?

7.2 The Conservation of Mechanical Energy

In a Nutshell

The **total mechanical energy** of a system is equal to the sum of the system's kinetic and potential energies.

The mechanical energy of a system is **conserved** if the total work done by all external forces and all internal nonconservative forces is zero.

Solving Problems Involving Mechanical Energy

Picture: Identify a system that includes the object(s) of interest and any other objects that interact with the object(s) of interest by either a conservative or kinetic-frictional force.

Solve:

1. Make a sketch of the system and include labels. Include coordinate axes and show the system in its initial and final configurations. Showing an intermediate configuration is often helpful also. Objects may be represented as dots, just as is done in free-body diagrams.

2. Identify any external forces acting on the system that do work, and any internal non-conservative forces that do work. Also identify any internal conservative forces that do work.

3. Apply the work-energy theorem for systems. For each internal conservative force doing work use a potential-energy function to represent the work done.

Check: Make sure that you have accounted for the work done by all conservative and non-conservative forces in determining your answer.

A particle is in an **equilibrium** position if the net force on it is zero. Because force is related to the derivative of the potential-energy function with respect to position, equilibrium corresponds to positions where $dU/dx = 0$. The equilibrium position can be either stable, unstable, or neutral depending on the curvature (second derivative) of the potential-energy function.

In **stable equilibrium**, a small displacement in any direction results in a restoring force that accelerates the particle back toward its equilibrium position. This corresponds to a position where the curvature of the potential-energy function is positive. An example of a stable equilibrium position is a ball sandwiched between two hills.

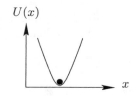

In **unstable equilibrium**, a small displacement results in a force that accelerates the particle away from its equilibrium position. This corresponds to a position where the curvature of the potential-energy function is negative. An example of an unstable equilibrium position is a ball at the top of a hill.

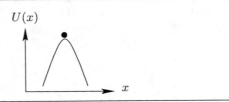

In **neutral equilibrium**, a small displacement in any direction results in zero force and the particle remains in equilibrium. This corresponds to a location where the curvature of the potential-energy function is zero.

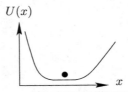

Fundamental Equations

Total mechanical energy

$$E_{\text{mech}} = K_{\text{sys}} + U_{\text{sys}}$$

Work-energy theorem for systems

$$W_{\text{ext}} = \Delta E_{\text{mech}} - W_{\text{nc}}$$

Conservation of mechanical energy

$$E_{\text{mech}} = K_{\text{sys}} + U_{\text{sys}} = \text{constant or } K_{\text{i}} + U_{\text{i}} = K_{\text{f}} + U_{\text{f}}$$

Force and potential energy

$$F_x = -\frac{dU}{dx} \ (\text{where } \vec{F} = F_x \hat{\imath} \text{ is conservative})$$

Stable equilibrium condition

$$\frac{d^2U}{dx^2} > 0$$

Unstable equilibrium condition

$$\frac{d^2U}{dx^2} < 0$$

Neutral equilibrium condition

$$\frac{d^2U}{dx^2} = 0$$

Common Pitfalls

> A system's total mechanical energy is not the same as a system's total energy. The total energy includes the total mechanical energy plus any additional forms of energy.
> Only *changes* in potential energy have any physical meaning. The choice of the reference point for potential energy is always arbitrary, so choose a reference configuration that is convenient for a particular situation.
> Although kinetic energy is always positive, potential energy can be either positive or negative.

3. TRUE or FALSE: External work refers to a process in which the system is displaced from an initial to a final configuration as a result of external forces that act on the system.

4. A block of mass m, released from rest, slides down a frictionless incline and reaches the bottom with speed v_{f}. What is the change in the total mechanical energy of the *block–slide–earth* system? Explain.

Try It Yourself #3

The potential-energy function for a conservative force $\vec{F} = F_x \hat{\imath}$ acting on a 2.00-kg particle is shown. For what values of x is the force (a) zero; (b) directed leftward; (c) directed rightward? (d) What values of x are equilibrium positions? (e) For each equilibrium position state whether it is a position of stable, neutral, or unstable equilibrium. (f) For what value(s) of x is the magnitude of the force the greatest?

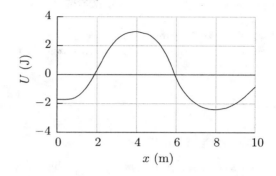

Picture: For a conservative force, $F_x = -dU/dx$, so we need to calculate (or at least estimate) the slope of the potential-energy function to answer questions about the force. Equilibrium conditions depend on whether or not the potential-energy function is at a maximum, minimum, or constant.

Solve:

(a) The force is zero where $dU/dx = 0$.	$0 \leq x \leq 0.7$ m, $x = 4$ m, $x = 8$ m
(b) The force is directed leftward, in the $-x$ direction, where $F_x < 0$, or $dU/dx > 0$.	0.7 m $< x < 4$ m, 8 m $< x < 10$ m
(c) The force is directed rightward, in the $+x$ direction, where $F_x > 0$, or $dU/dx < 0$.	4 m $< x < 8$ m
(d) Equilibrium positions are positions at which the force is zero.	$0 \leq x \leq 0.7$ m, $x = 4$ m, $x = 8$ m
(e) A stable equilibrium occurs at a potential-energy minimum, an unstable equilibrium occurs at a potential-energy maximum, and a neutral equilibrium occurs where the potential-energy function is constant.	neutral, unstable, and stable equilibria, respectively
(f) The magnitude of the force is greatest where the slope of the potential-energy function is the steepest.	$x \approx 2$ m, $x \approx 5.8$ m

Check: Make sure you know your derivatives.

Taking It Further: If a particle is moving under the influence of only this force, approximately where would the power delivered to the particle be the greatest? Why?

Try It Yourself #4

Blocks of mass m_1 and $m_2 < m_1$ are hung over a massless, frictionless pulley using a taut string of negligible mass, as shown. When the system is released from rest, block 2 is in contact with the floor. Following release, block 1 falls a distance h to the floor as block 2 is pulled upward through the same distance. To what maximum height does block 2 rise above its starting position if $m_1 = 6.00$ kg, $m_2 = 5.00$ kg, and $h = 1.50$ m?

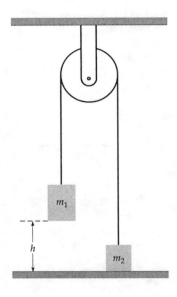

Picture: You might initially think that block 2 will simply rise a distance h. However, this is not the case because when block 2 reaches this height, it will have some velocity that will continue to carry it upward an additional distance. This problem consists of one segment (A) while block 1 is falling, and a second segment (B) during which block 1 remains on the floor and block 2 continues to rise. For segment A, choose the system to be the two blocks, the string, the pulley, the pulley mount, the ceiling, and Earth, so that the mechanical energy of the system is conserved. For segment B, consider the system to consist of only block 2 and Earth, again allowing the conservation of mechanical energy.

Solve:

1. Draw three pictures of the masses on the pulley. The first one should show their initial positions. The second should show their positions just before mass 1 hits the ground. This is the "final" configuration for the segment A, and the "initial" configuration for segment B. Finally, draw the system with mass 2 at its highest position, which is the "final" configuration for the segment B. Label the drawings as appropriate.   2. Place a coordinate system on the drawing. Choose one with the $+x$ direction pointed to the right and the $+y$ direction up. For convenience, let the floor be the position $y = 0$.   3. Near each picture, draw a vector representing the velocity at the time shown. Label the vector. If you know the initial and/or final velocities, include that information on the pictures. Also include whatever position (height) information you know for each configuration.

Determine the initial energy of the system for segment A. Include the potential and kinetic energies associate with each mass.	
Find an *algebraic* expression for the final energy of the system for segment A.	
Set the initial and final energies of segment A equal to each other to find an *algebraic* expression for the speed of block 2 at the end of segment A.	
Find an *algebraic* expression for the initial energy of the much smaller system for segment B. The final velocity of mass 2 for segment A is the initial velocity of mass 2 for segment B.	
Find an *algebraic* expression for the final energy of the system for segment B.	

Set the initial and final energies of segment B equal to each other and solve for the final height of mass 2.	
	$h_2 = 1.64$ m

Check: As expected, we found that the second mass does reach a greater height than the initial height of mass 1.

Taking It Further: If you consider just the work done by the gravitational force on the two masses, does the force of gravity do the same amount of work on each mass? Why or why not?

7.3   The Conservation of Energy

In a Nutshell

The **law of conservation of energy** states that the total energy of the universe is constant. Energy can be converted from one form to another, or transmitted from one region to another, but energy can never be created or destroyed.

The energy of a system can change if a force does work on it. However, if thermal energy is added to or removed from the system its energy will also change. Some possible sources of heat include frictional forces, chemical reactions, and the emission or absorption of electromagnetic radiation.

When surfaces slide across each other, kinetic friction decreases the mechanical energy of the system and increases the thermal energy.

Fundamental Equations

Law of conservation of energy $\quad E_{\text{in}} - E_{\text{out}} = \Delta E_{\text{sys}}$

Energy of a system $\quad E_{\text{sys}} = E_{\text{mech}} + E_{\text{therm}} + E_{\text{chem}} + E_{\text{other}}$

Work-energy theorem $\quad W_{\text{ext}} = \Delta E_{\text{sys}} = \Delta E_{\text{mech}} + \Delta E_{\text{therm}} + \Delta E_{\text{chem}} + \Delta E_{\text{other}}$

Energy dissipated by kinetic friction $\quad f_{\text{k}} s_{\text{rel}} = \Delta E_{\text{therm}}$

Common Pitfalls

> ➤ Don't confuse a system's total mechanical energy with its total energy. The total energy includes the total mechanical energy and any additional forms of energy.
> ➤ Keep in mind that the amount of mechanical energy that is transformed into thermal energy when two surfaces slide across each other is $f_k s_{rel}$, where f_k is the kinetic frictional force and s_{rel} is the distance one surface moves relative to the other surface. Don't mistake $f_k s_{rel}$ for the work done by the kinetic frictional force. It isn't. The work done by f_k cannot be directly calculated because the displacements of the points where this force is applied are not directly observable.

5. TRUE or FALSE: A block sliding on a horizontal floor is brought to rest by a kinetic frictional force. All of the dissipated kinetic energy appears as the thermal energy of the block.

6. A boy pulls a wagon up a long, straight hill at constant speed. Does the total energy of the *wagon–hill–earth* system change? Explain.

Try It Yourself #5

You are inside a crate that is released from rest at the top of a ramp inclined 30° with the horizontal. After sliding a distance $L = 4.00$ m down the ramp the crate runs into a spring bumper which it compresses as it slows. You and the crate have a mass of 80.0 kg, the spring is massless and has a spring constant of 500 N/m, and the coefficient of kinetic friction between the block and the ramp is 0.300. What is the maximum distance that the spring is compressed?

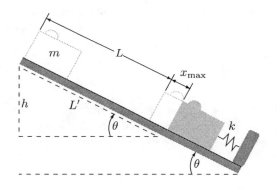

Picture: A sketch of the system is shown. The points of interest are when you are initially at the top of the ramp and at the bottom of the ramp when you have fully compressed the spring. If the system consists of you, the crate, the ramp, the spring, and Earth, the initial potential energy of the system will be converted into thermal energy through friction, and into the elastic potential energy stored in the compressed spring. There are no external forces acting on the system, so the total energy of the system remains unchanged.

Solve:

Write the work-energy theorem in generic *algebraic* form as a guide for solving the problem.	

Determine *algebraically* the work done by external forces. Hint: How many external forces are there?	
Determine *algebraically* the initial, final, and change in the kinetic energy. You and the crate are at rest both initially and at the maximum compression of the spring.	
Determine *algebraically* the initial, final, and change in the gravitational potential energy. From the labeling in the figure, you and the crate start out at height h and end at a height referenced as zero.	
Determine *algebraically* the initial, final, and change in the elastic potential energy of the spring. The spring is initially uncompressed and at the end has a maximum compression.	
Determine *algebraically* the amount of mechanical energy converted to heat due to kinetic friction. Remember to include the entire distance the crate slides.	
Draw a free-body diagram of you and the crate combined, and use Newton's second law to find the normal force. You need this and the coefficient of kinetic friction to find the frictional force.	

Substitute all results so far (still algebraically) into the work-energy theorem from the first step. Relate the height h to the distance $L + x_{max}$, using their geometric relationship.	
This is a quadratic equation in x_{max}. Solve it by rearranging it into standard form (still algebraically).	
Finally, substitute in numbers. Since x_{max} is a distance, not a coordinate, we need to use the positive solution of our quadratic equation.	$x_{max} = 2.16$ m

Check: Make sure your units agree throughout this entire calculation.

Taking It Further: If the coefficient of kinetic friction were increased, would the spring compress more, less, or the same amount? Why?

Try It Yourself #6

Starting from rest, Buck the sled dog drags a 45.0-kg sled up a 5.00-m long ramp that is inclined 35° with the horizontal. The kinetic coefficient of friction between the ramp and the sled is 0.350. Using the work-energy theorem, determine how much work Buck must do on the sled just to drag it up the ramp. Assume that Buck's paws do not slip on the ramp, and that the initial and final velocities of the sled are zero.

Picture: The points of interest are when the sled is initially at rest at the bottom of the ramp and when the sled has reached the top of the ramp. If the system consists of the sled, the ramp, and the earth, the force of Buck's pulling will be an external force that does work on the system. The work done by Buck can be converted into gravitational potential energy, kinetic energy, and thermal energy through friction.

Solve:

Draw a sketch of the problem, identifying the important points.	
Write the work-energy theorem in a general, algebraic form as a guide for solving the problem.	
Determine an *algebraic* expression for the work done by external forces, in this case the work done by Buck.	
Determine, *algebraically*, the initial, final, and change in the kinetic energy.	
Determine, *algebraically*, the initial, final, and change in the gravitational potential energy.	

Determine, *algebraically*, the amount of mechanical energy converted to heat as a result of the frictional force.	
Draw a free-body diagram of the sled and use Newton's second law to find the normal force.	
Substitute everything into the work-energy theorem, still algebraically.	
Solve algebraically for the work done by Buck. Then substitute numerical values.	$W = 1.90 \times 10^3$ J

Check: We expect the work done by Buck to be positive because he has to provide the energy for the increase in the gravitational potential energy of the system.

Taking It Further: How can the sled slide on the ice when Buck's paws do not?

7.4   Mass and Energy

In a Nutshell

Any material particle has an intrinsic energy called its rest energy E_0. A particle's rest energy is given by the famous equation $E_0 = mc^2$.

Fundamental particles combine to form atomic nuclei, so we can treat an atomic nucleus as a system of these particles. Since the particles that are bound together in a nucleus attract one another, energy must be supplied to separate them. The minimum energy required to separate a nucleus into its constituent particles is called its binding energy. Atomic and nuclear energies are usually measured in units called electron volts (eV).

Energy is equivalent with mass. Since the energy of a bound nucleus is less than the energy when the same particles are widely separated, the mass of the system in the nucleus, with its particles bound together, is less than the sum of the masses of the widely separated constituent particles. This difference in energy is referred to as the **binding energy** of the nucleus or other structure.

Fundamental Equations

Rest energy
$$E_0 = mc^2$$

Important Derived Results

Relativistic kinetic energy
$$K = \tfrac{1}{2} E_0 \frac{v^2}{c^2}$$

Common Pitfalls

➢ The mass of a system of particles is *not* always equal to the sum of the masses of constituent fundamental particles. If the system is bound, then the energy is less than it would be if the particles are widely separated. Since the mass of a system is proportional to its energy, a bound system has less mass than a system of the same particles if they are widely separated. The difference is called the binding energy. These distinctions are always valid, but they are most significant for subatomic systems.

7.5   Quantization of Energy

In a Nutshell

For bound systems, like an atomic nucleus or an atom, the total energy of the system is found to be **quantized**. That is, the smallest increases or decreases of a bound system's total energy occur in finite amounts, so the system's energy levels are discrete (quantized). Because these energy changes are small in comparison to the changes that typically occur in our everyday life, the finite size of the smallest increments in energy were not noticed until about a century ago.

Electromagnetic energy (radio waves, microwaves, light waves, X-rays, and gamma rays) is always absorbed or emitted by an amount directly proportional to the frequency of the radiation.

Physical Quantities and Their Units

Planck's constant
$$h = 6.626 \times 10^{-34} \text{ J} \cdot \text{s} = 4.136 \times 10^{-15} \text{ eV} \cdot \text{s}$$

Fundamental Equations

Energy of a photon
$$E = hf$$

QUIZ

1. TRUE or FALSE: The work done by a kinetic frictional force f_k equals $f_k s_{rel}$, where s_{rel} is the distance one surface slides relative to the other surface.

2. TRUE or FALSE: The initial and final energies of a system are the sums of all types of energy (kinetic, potential, chemical, thermal, ...) at two specific times, the initial and final times.

3. A boy pulls a wagon up a long, straight hill at constant speed. What is the change in total energy of the *boy–wagon–hill–earth* system? Explain.

4. When you step on the accelerator, your car's kinetic energy increases. What external force, if any, does work on the car to cause this increase in its kinetic energy? Where does this kinetic energy come from?

5. A blob of putty falls on the floor (plop). Neglecting air resistance, does the total energy of the *putty–floor–earth* system change? Explain.

6. Blocks of mass m_1 and m_2 are hung over a mass-less, frictionless pulley by a light string, as shown. There is no friction between the block and the incline. When the system is released from rest in the position shown, block 1 descends a distance h to the ground as the string pulls block 2 along the incline through the same distance. What is the maximum total displacement along the incline that block 2 undergoes when $m_1 = m_2 = 5.00$ kg, $h = 1.00$ m, and $\theta = 30°$?

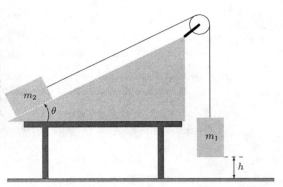

7. A pendulum consists of a compact particle of mass m suspended from a string of length L attached to a mount on the ceiling, as shown. The particle is released from rest with the string horizontal. As the particle passes through the lowest point in its path, the string strikes a skinny peg a distance bL above the lowest point. As the particle passes through point P, a distance bL directly above the peg, determine an expression for the tension in the string in terms of m, g, L, and b. Find the value of b in the limit that the tension in the string approaches zero. Assume air drag and friction are negligible. For certain values of b, the equation yields a negative value for the tension. Is that physically possible? Explain.

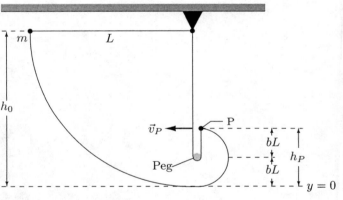

Chapter 8

Conservation of Linear Momentum

8.1   Conservation of Linear Momentum

In a Nutshell

The **linear momentum** of a particle is defined as the product of its mass and its velocity. Since velocity is a vector quantity, linear momentum is also a vector quantity.

Newton's second law can be expressed in terms of momentum: The net force acting on a particle is equal to the time rate of change of the particle's linear momentum.

The **law of conservation of momentum** states that If the sum of the external forces on a system remains zero, the total momentum of the system remains constant. Because momentum is a vector quantity, the law of conservation of momentum can be applied to each component of the momentum individually. For instance, it is possible to conserve the horizontal component of the momentum even if the vertical component of momentum is not conserved.

Finding Velocities Using Momentum Conservation

Picture: Determine that the net external force $\sum \vec{F}_{\text{ext}}$ (or at least one component of the net external force) on the system is negligible for some interval of time. If the net force is determined not to be negligible, do not proceed.

Solve:

1. Draw a sketch showing the system both before and after the time interval. Include coordinate axes and label the initial and final velocity vectors.
2. Equate the initial momentum to the final momentum. That is, write the equation $m_1\vec{v}_{1\text{i}} + m_2\vec{v}_{2\text{i}} = m_1\vec{v}_{1\text{f}} + m_2\vec{v}_{2\text{f}}$ (or $m_1 v_{1\text{i}x} + m_2 v_{2\text{i}x} = m_1 v_{1\text{f}x} + m_2 v_{2\text{f}x}$).
3. Substitute the given information into the step-2 equation(s) and solve for the quantity of interest.

Check: Make sure you include any negative signs that accompany velocity components because they influence your final answer.

Physical Quantities and Their Units

momentum $\vec{p}$ dimensions of mass times velocity or units of $\text{kg} \cdot \text{m/s}$

Fundamental Equations

Linear momentum of a particle $\vec{p} = m\vec{v}$

Newton's second law $\vec{F}_{\text{net}} = \dfrac{d\vec{p}}{dt}$

Important Derived Results

Linear momentum of a system
$$\vec{P}_{\text{sys}} = \sum_i m_i \vec{v}_i = M \vec{v}_{\text{cm}}$$

Conservation of momentum
$$\text{If } \sum \vec{F}_{\text{ext}} = 0, \text{ then } \vec{P}_{\text{sys}} = \text{constant}$$

Common Pitfalls

> Just because momentum is not conserved in one direction does *not* mean it is not conserved in any direction. The law of conservation of momentum is a vector statement. If the component of the net external force on a system in a given direction is zero, the component of the system's momentum in that direction remains constant.

1. TRUE or FALSE: If the sum of the internal forces in a system remains zero, the momentum of the system necessarily remains constant.

2. When a spaceship in empty gravity-free space fires a rocket, does the momentum of the ship–fuel–rocket system change? Does the momentum of the ship change? Does the momentum of the rocket change?

Try It Yourself #1

A 3.00-kg rifle, directed horizontally and initially at rest, fires a bullet of mass 10.0 g at a muzzle speed of 650 m/s. Assuming the rifle is free to move, at what speed would it recoil?

Picture: The momentum of the rifle–bullet system is conserved in the horizontal direction. In this case the momentum is equal to zero before and after firing the rifle.

Solve:

Write an algebraic expression for conservation of momentum in this case. What do you know about the initial momentum? Make sure to include the momentum of both the bullet and the rifle.	
Algebraically solve for the recoil speed of the rifle and substitute numerical values and their units. If the bullet is fired in the positive direction, the rifle velocity should be in the negative direction.	$v_{\text{f}} = 2.17$ m/s

Check: The units are correct, and the final momentum should add to zero.

Taking It Further: Why is it generally important for the butt of the rifle to be firmly supported when firing?

Try It Yourself #2

A 900-kg car traveling north at 60.0 km/h collides with a 1200-kg light truck traveling west at 50.0 km/h. The vehicles stick together following the collision, as shown. Determine the velocity (magnitude and direction) of the wreck immediately following the collision. Neglect friction between the vehicles and the ground.

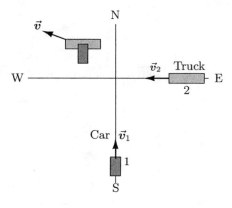

Picture: The momentum of the car–truck system before and after the collision must be the same. This is a two-dimensional problem, so momentum must be conserved in both the x and y directions. Let the positive x direction be east and the positive y direction be north.

Solve:

Write the conservation-of-momentum equation, algebraically and in vector form, as a guide for the rest of the problem.	
Write the equation for conservation of momentum in the x direction, using algebraic representations.	
Rearrange the above expression to solve for the final x component of the velocity, then substitute in values to find v_{xf}. Because the vehicles stick together, they both have the same final velocity.	

Write the equation for conservation of momentum in the y direction, using algebraic representations.	
Rearrange the above expression to solve for the final y component of the velocity, then substitute in values to find v_{yf}.	
Find the magnitude and direction of the final velocity.	
	$v = 38.4$ km/h at $42.0°$ north of west

Check: Make sure your units are correct. One vehicle was traveling north, and the other west before the collision. As a result, we would expect them to be traveling in a generally northwest direction after the collision.

Taking It Further: How would the solution to this problem change if the cars did not stick together?

8.2   Kinetic Energy of a System

In a Nutshell

The kinetic energy of a system can be expressed as the sum of two terms: the kinetic energy associated with the center-of-mass motion and the kinetic energy associated with the motions relative to the center of mass.

Important Derived Results

Kinetic energy of a system of particles
$$K = \sum_i \tfrac{1}{2} m_i v_{cm}^2 + \sum_i \tfrac{1}{2} m_i u_i^2 = \tfrac{1}{2} M v_{cm}^2 + K_{rel}$$

8.3 Collisions

In a Nutshell

In an **elastic collision** the total kinetic energy of the system is conserved.

In one-dimensional (head-on) elastic collisions, the speed of separation equals the speed of approach. This is a consequence of the conservation of *both* kinetic energy *and* momentum. In the collision depicted in the figure, the velocity of particle 1 changes sign as a result of the collision.

Speed of approach $= v_1 - v_2$

Speed of separation $= v_2 + v_1$

In an **inelastic collision** the total kinetic energy of the system is *not* conserved.

In a **perfectly inelastic collision** all kinetic energy relative to the center of mass is converted to thermal or internal energy of the system, and the objects share a common final velocity usually because they stick together.

The **coefficient of restitution** e is a measure of the elasticity of a collision. For elastic collisions $e = 1$. For perfectly inelastic collisions $e = 0$.

Forces transfer momentum to the objects they act on. The measure of the momentum transferred by a force is the **impulse** $\vec{I}$. It is defined as the time integral of the force acting on the object. The net **impulse** on a particle is equal to the particle's change in momentum.

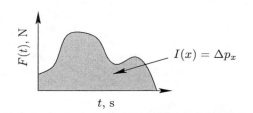

Estimating the Average Force

Picture: To estimate the average force $\vec{F}_{av}$ we first estimate the impulse of the force $\vec{I}$. The impulse of the force equals the net impulse assuming any other forces are negligible. The net impulse is equal to the change in momentum, and the change in momentum equals the product of the mass m and the change of velocity $\vec{v}_f - \vec{v}_i$. An estimate of the change in velocity can be gotten from estimates of both the collision time Δt and the displacement $\Delta \vec{r}$.

Solve:

1. Calculate or estimate the impulse $\vec{I}$ and the time Δt. This estimate assumes that during the collision, the collision force on the object is large compared to all other forces on the object. This procedure works *only* if the displacement during the collision can be determined.
2. Draw a sketch showing the before and after position of the object. Add coordinate axes and label the pre- and post-collision velocities $\vec{v}_i$ and $\vec{v}_f$. In addition, label the displacement $\Delta \vec{r}$ during the collision.
3. Calculate the change in momentum of the object during a collision. The impulse on the object equals its change in momentum ($\vec{I} = \Delta \vec{p} = m \Delta \vec{v}$).
4. Use kinematics to estimate the collision time. This means using both $\vec{v}_{av} \approx \frac{1}{2}(\vec{v}_i + \vec{v}_f)$ and $\Delta \vec{r} = \vec{v}_{av} \Delta t$ to obtain $\Delta \vec{r} \approx \frac{1}{2}(\vec{v}_i + \vec{v}_f)\Delta t$, and then solving for Δt.
5. Use $\vec{F}_{av} = \vec{I}/\Delta t = m \Delta \vec{v}/\Delta t$ to calculate the average force.

Check: Average force is a vector. Your result for average force should have the same direction as the change in velocity vector.

Physical Quantities and Their Units

Impulse $\vec{I}$ dimensions of force×time, or units of N · s

Fundamental Equations

Impulse
$$\vec{I} = \int_{t_1}^{t_2} \vec{F}\, dt$$

Impulse–momentum theorem
$$\vec{I}_{\text{net}} = \int_{t_1}^{t_2} \vec{F}_{\text{net}}\, dt = \Delta\vec{p}$$

Important Derived Results

Average force
$$\vec{F}_{\text{av}} = \frac{1}{\Delta t}\int_{t_1}^{t_2} \vec{F}\, dt = \frac{\vec{I}}{\Delta t}$$

Relative speeds of approach and separation
$$v_{2f} - v_{1f} = v_{1i} - v_{2i}$$

Coefficient of restitution
$$e = \frac{v_{2f} - v_{1f}}{v_{1i} - v_{2i}}$$

Common Pitfalls

> It is often the case that during collisions, impulses associated with external forces acting on the system are negligible compared to impulses associated with the internal forces between the colliding objects. Thus even for systems that are not isolated, it is often a good approximation to equate the system's pre- and post-collision momenta if the collision occurs over a sufficiently short time interval. During a collision, the momentum of the system remains constant to the degree that impulses associated with external forces are negligible.

> When a perfectly inelastic one-dimensional collision occurs, the kinetic energy of the center-of-mass motion remains constant while the kinetic energy relative to the center of mass becomes zero. Thus the final kinetic energy of the system cannot be zero unless the velocity of the center of mass is zero.

> Remember that if a collision is completely elastic, momentum conservation is not sufficient to solve the problem. You will also have to use conservation of kinetic energy.

> When determining impulse, remember it is the force integrated over time, *not* distance.

3. TRUE or FALSE: Following a head-on, perfectly inelastic collision, the kinetic energy of the colliding particles necessarily equals zero.

4. The average force required to change the momentum of a system by a specific amount is less the longer the time interval in which the change takes place. Does a longer time interval also mean that a smaller maximum force is required?

Try It Yourself #3

A 2.00-kg sphere with a velocity of $(8.00 \text{ m/s})\hat{\imath}$ runs into a stationary 8.00-kg sphere. Following the collision the 2.00-kg sphere moves with a velocity of $(3.00 \text{ m/s})\hat{\imath} - (4.00 \text{ m/s})\hat{\jmath}$. Determine the velocity of the 8.00-kg sphere following the collision.

Picture: Sketch the motion of the spheres both before and after the collision. The momentum of the system is conserved in both the x and y directions.

Solve:

Sketch the motion of the spheres both before and after the collision	
Write *algebraic* expressions for the law of conservation of momentum, in both vector and component form.	
Substitute values into the x component equation.	
Substitute values into the y component equation.	
Simultaneously solve the resulting equations from the previous two steps to find the final velocity of the 8.00-kg sphere.	$\vec{v}_{\text{f,8}} = (1.25 \text{ m/s})\hat{\imath} + (1.00 \text{ m/s})\hat{\jmath}$

Check: The units work out, and the general direction of motion of the center of mass appears to be in the $+x$ direction, which was the direction of the initial motion of the system.

Taking It Further: Is this collision elastic or inelastic? How can you tell?

Try It Yourself #4

An empty, open railroad car of mass M with frictionless wheels is rolling along a horizontal track with speed v_0 when it begins to rain. There is no wind so the rain falls straight down. The mass of the rainwater that accumulates in the car is m. (a) Determine the speed of the car after the rain has accumulated. (b) After the rain stops, but while the car is still moving, a worker opens a drain hole in the bottom of the car and lets the water out. Determine the speed of the car when it is once again empty.

Picture: This is a conservation of momentum problem in the horizontal direction, because the only external forces, weight and a normal force, are directed in the vertical direction. The collision between the car and the rainwater is perfectly inelastic. The rain initially has no horizontal momentum, but it eventually acquires some. For the second "collision," when the rainwater is let out, think carefully about the horizontal velocity of the water the instant it leaves the rail car.

Solve:

Draw a sketch illustrating the motion of the car and rainwater before and after each collision.	
Write the expression for the conservation of momentum in the horizontal direction in *algebraic* form.	
Rearrange the expression to solve for the final velocity of the railroad car. What do you know about the initial and final horizontal velocities of the rain?	$$v = \frac{M}{m + M} v_0$$

Conserve momentum again while the water is being let out. What do you know about the initial and final velocities of the water just before and just after the water is let out?	
	$$v = \frac{M}{m + M} v_0$$

Check: The railroad car slows down to compensate for the fact that the rain has gained some horizontal momentum.

Taking It Further: Opening up the drain hole does not allow the speed of the train car to increase because the water still has the same horizontal momentum the instant after it is drained. Is there any way for the railroad worker to increase the speed of the car again without attaching it to an engine that provides an external force?

8.4   Collisions in the Center-of-Mass Reference Frame*

In a Nutshell

Any reference frame in which the center of mass of a system remains stationary is called a **center-of-mass reference frame** for that system. In this frame the total momentum of the system, which equals the product of the total mass of the system and the velocity of the center of mass, also equals zero. Consequently, a **center-of-mass reference frame** is also a **zero-momentum reference frame**.

In the center-of-mass reference frame, the momenta of the two incoming objects just before a collision must be equal and opposite in order for the momentum to be zero.

Important Derived Results

Velocity relative to center of mass $\vec{u}_1 = \vec{v}_1 - \vec{v}_{\text{cm}}$

---

*Optional material.

Common Pitfalls

> ➤ Remember that in the center-of-mass reference frame the total momentum is always zero.

5. TRUE or FALSE: The total kinetic energy of a system in the center-of-mass reference frame is always zero.

6. Is it possible for an object to have a velocity $\vec{v}$ in one reference frame, and a velocity of zero in another reference frame? Why or why not?

Try It Yourself #5

Consider a head-on elastic collision be-tween a cue ball and the eight ball. As-suming the cue ball has an initial veloc-ity $\vec{v}_0$, the eight ball is initially rest, and the balls have equal masses m, determine the velocities of each ball immediately fol-lowing impact. To determine these veloci-ties, transfer to the center-of-mass system, determine the post-collision velocities in that frame of reference, and then transfer back to the initial frame of reference.

Picture: Determine the velocity of the center of mass. Remember that in a one-dimensional elastic collision, like this one with two spheres, the velocities relative to the center of mass get reversed. The figure helps to illustrate this.

Solve:

Find the velocity of the system's center of mass.	
Find the initial velocities of the two balls rela-tive to the center of mass velocity.	

For elastic, one-dimensional collisions, the speeds relative to the center of mass get reversed, and the speed of the center of mass remains constant. We can use this information to find the speeds of the balls after the collision.	
	$v_{2,\mathrm{f}} = v_0,\ v_{1,\mathrm{f}} = 0\ \mathrm{m/s}$

Check: The eight ball after the collision has all the momentum of the cue ball before the collision. Momentum is conserved.

Taking It Further: How would the problem change if the eight ball had twice the mass of the cue ball?

8.5   Continuously Varying Mass and Rocket Propulsion

In a Nutshell

Conservation of momentum can be applied to systems whose mass changes with time. Such a system might include a rocket ship and the unspent fuel in it, or an object flying through the air that picks up contaminants as it moves.

Important Derived Results

Newton's second law for continuously variable mass

$$\vec{F}_{\mathrm{net\ ext}} + \frac{dM}{dt}\,\vec{v}_{\mathrm{rel}} = M\frac{d\vec{v}}{dt}$$

Try It Yourself #6

Comets grow in size by collecting galactic dust as they travel through the solar system. Let a comet have an initial mass M and an initial speed v_{i} in the "lab" frame of reference, and be collecting dust (adding mass) at a rate of R. Furthermore, for simplicity let the galactic dust be stationary in this frame of reference, and assume there are no additional forces on the growing comet. Find an expression for the speed of the comet as a function of time.

Picture: We will have to use Newton's second law for a continuously varying mass.

Solve:

Write the algebraic expression for Newton's second law for a continuously varying mass.	
Determine the velocity of the dust relative to the rest of the comet.	
Substitute the values for $\vec{F}_{\text{net ext}}$, dM/dt, and $\vec{v}_{\text{rel}}$ for this particular problem into Newton's second law.	
Re-arrange the expression to get dv/v on one side and all other terms on the other side.	
Integrate both sides of the equation. The "t" side will be integrated from $t = 0$ to some time t_{f}. The "v" side will be integrated from v_{i} to v_{f}. Then take e to the power of both sides, and re-arrange to find an expression for v_{final}.	$v_{\text{f}} = v_{\text{i}}e^{-(R/M)t_{\text{f}}}$

Check: In the absence of other forces, we should expect the speed of the comet to be reduced. That is the only way momentum can be conserved. Our answer demonstrates that characteristic.

Taking It Further: What would be required to keep the comet traveling at a constant speed?

QUIZ

1. TRUE or FALSE: The total momentum of a system is the product of its total mass and the velocity of its center of mass.

2. TRUE or FALSE: The impulse associated with a force is a measure of the mechanical energy transferred.

3. Can a system have zero kinetic energy and nonzero momentum? Can it have zero momentum and nonzero kinetic energy? Explain.

4. When a large truck collides head on with a small car, the force exerted by the car on the truck is equal in magnitude and oppositely directed to the force exerted by the truck on the car. Would the momentum of the car–truck system be conserved if this were not so? Explain.

5. Is momentum always conserved during a collision? Why or why not?

6. Rainwater is falling straight down with a velocity of 15.0 m/s at the rate of 10.0 cm of rain per hour. An electronic scale is placed in the rain. The scale's pan consists of a 25.0-cm-diameter circular disk. If the weight of the water on the platform is negligible, determine the reading of the scale.

7. A cue ball moving with a speed of v_0 glances off the initially stationary two ball. Both balls have the same mass. The collision is perfectly elastic, and the cue ball is deflected 30° from its pre-collision path. What is the velocity of the two ball following the collision?

Chapter 9

Rotation

9.1 Rotational Kinematics: Angular Velocity and Angular Acceleration

In a Nutshell

Suppose a disk is constrained to rotate about a fixed axis through its center and perpendicular to the disk as shown. This is considered one-dimensional rotational motion. Either the clockwise or counterclockwise direction of rotation can be considered the "positive" direction. A radial line drawn on the disk will sweep out angle θ as the disk rotates. This angle between the drawn line and a reference direction is called the **angular position** of the disk. The change in the angular position $d\theta$ is called the **angular displacement**, and the instantaneous rate of change of the angular position, $\omega = d\theta/dt$, is the **angular velocity**. The instantaneous rate of change of the angular velocity, $\alpha = d\omega/dt$, is the **angular acceleration**.

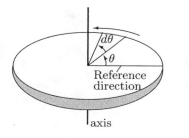

The common units of angle are the degree, the radian, and the revolution. The radian is the SI unit of angle. The radian is dimensionless because it is defined as the ratio of two lengths. By definition, 1 radian (rad) is the measure of the central angle of a circle whose intercepted arc length equals the radius.

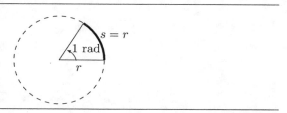

Rotational motion about a fixed axis is directly analogous to one-dimensional motion, and the relationship between their kinematic expressions are as follows:

Quantity	Linear Motion	Rotational Motion
Position	x	θ
Displacement	Δx	$\Delta \theta$
Average velocity	$v_{av} = \dfrac{\Delta x}{\Delta t}$	$\omega_{av} = \dfrac{\Delta \theta}{\Delta t}$
Instantaneous velocity	$v = \dfrac{dx}{dt}$	$\omega = \dfrac{d\theta}{dt}$
Average acceleration	$a_{av} = \dfrac{\Delta v}{\Delta t}$	$\alpha_{av} = \dfrac{\Delta \omega}{\Delta t}$
Instantaneous acceleration	$a = \dfrac{dv}{dt} = \dfrac{d^2 x}{dt^2}$	$\alpha = \dfrac{d\omega}{dt} = \dfrac{d^2\theta}{dt^2}$

Physical Quantities and Their Units

Angular position θ	dimensionless, but can have units of degrees (°), radians (rad), or revolutions (rev)
Angular velocity ω	dimensions of T^{-1}, with units of °/s, rad/s, rev/s, rev/min = RPM
Angular acceleration α	dimensions of T^{-2}, with units of °/s^2, rad/s^2, rev/s^2

Fundamental Equations

Average angular velocity	$\omega_{av} = \dfrac{\Delta \theta}{\Delta t}$
Instantaneous angular velocity	$\omega = \dfrac{d\theta}{dt}$
Average angular acceleration	$\alpha_{av} = \dfrac{\Delta \omega}{\Delta t}$
Instantaneous angular acceleration	$\alpha = \dfrac{d\omega}{dt} = \dfrac{d^2\theta}{dt^2}$
Angular conversion factors	$\dfrac{2\pi \text{ rad}}{1 \text{ rev}}, \dfrac{1 \text{ rev}}{360°}, \dfrac{\pi \text{ rad}}{180°}$

Important Derived Results

Kinematic expressions for constant α

$$\omega_{\mathrm{av}} = \tfrac{1}{2}(\omega_1 + \omega_2)$$

$$\omega = \omega_0 + \alpha(\Delta t)$$

$$\Delta\theta = \theta - \theta_0 = \omega_0(\Delta t) + \tfrac{1}{2}\alpha(\Delta t)^2$$

$$\omega^2 = \omega_0^2 + 2\alpha(\Delta\theta)$$

Relations between linear and angular kinematic parameters

$$\Delta s = r_i\,\Delta\theta$$

$$v_{\mathrm{t}} = r_i\omega$$

$$a_{\mathrm{t}} = r_i\alpha$$

$$a_{\mathrm{c}} = \frac{v^2}{r_i} = r_i\omega^2$$

Common Pitfalls

➤ What is a radian? If you cut a wedge of pumpkin pie and the length of one side of the wedge equals the length of the crust, the wedge angle equals 1 radian.

➤ Angular measures do not always have to be in radians in all the kinematic equations. Equations with *only* rotational kinematic parameters (e.g., θ, ω, and α) and time, such as $\Delta\theta = \omega(\Delta t)$, are valid with *any* consistent unit of angular measure. Thus, in the kinematic equations for rotational motion with constant α any consistent unit for angular measure may be used. Equations with *both* linear and angular parameters such as $\Delta s = r(\Delta\theta)$ are valid *only* when the unit of angular measure is the radian.

➤ Don't be misled into thinking that linear acceleration means the acceleration component in the direction of the velocity vector. In general, linear acceleration $\vec{a}$ has both a centripetal, a_{c}, and a tangential, a_{t}, component. Linear position, velocity, and acceleration are the same old friends that we called position, velocity, and acceleration before introducing the corresponding angular terms. When referring to angular position, angular velocity, or angular acceleration, it is best to explicitly use the "angular" descriptor.

➤ For a rotating object, $r\omega$ is the linear velocity of a point a distance r from the rotation axis. Don't fall into the trap of thinking that for a rotating object $r\alpha$ is the linear acceleration of the point. This is only the tangential component of the linear acceleration vector. There is also a centripetal component given by $r\omega^2$.

1. TRUE OR FALSE: All parts of a rotating rigid object have the same angular velocity.

2. Consider two points on a disk rotating with increasing speed about its axis, one at the rim and the other halfway from the rim to the center. Which point has the greater (a) angular acceleration, (b) tangential acceleration, (c) radial acceleration, (d) centripetal acceleration?

Try It Yourself #1

A phonograph turntable with a mass of 1.80 kg and a radius of 15.0 cm is being braked to a stop. After 5.00 s its initial angular speed has decreased by 15%. Assuming constant angular acceleration, (a) how long does it take to come to rest? (b) If the initial angular speed is 33-1/3 RPM, through how many revolutions does it turn while coming to rest?

Picture: This is a kinematics problem with constant angular acceleration. We are not given an initial angular speed, so we must assume part (a) can be solved without it. Assign the initial angular speed the symbol ω_0.

Solve:

Draw a sketch to help visualize the situation	
Convert the angular speed to rev/s to avoid confusing yourself later.	
(a) Write two constant-angular-acceleration equations. One will describe the fact that 85% of the angular speed remains after 5.00 s. The other will describe the fact that after some time t, the angular speed is zero. By dividing these equations, you can solve for t.	$t = 33.3$ s
(b) Write a constant angular acceleration equation for the position. Assume the initial position is zero. Knowing ω_0 and t from the previous part, you can find the angular acceleration and then the final angular position.	9.26 rev

Check: All the units work out. You can use the angular unit conversion factors if you solved the problem with different angular units.

Taking It Further: Does a constant angular acceleration correspond to a constant tangential acceleration, constant centripetal acceleration, neither, or both? Explain.

Try It Yourself #2

Point P is located on a record turntable a distance of 10.0 cm from the axis. A dime is placed on the turntable directly over point P as shown. The coefficient of static friction between the dime and the turntable is 0.210. If the turntable starts from rest with a constant angular acceleration of 1.20 rad/s^2, how much time passes before the dime begins to slip?

Picture: The linear acceleration of P has both a tangential and a centripetal component. The tangential component is constant, but the centripetal component increases as the speed increases. The dime will slip when the frictional force can no longer provide the same acceleration to the dime that point P experiences.

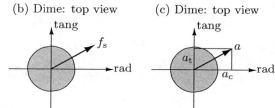

(b) Dime: top view

(c) Dime: top view

(*a*)

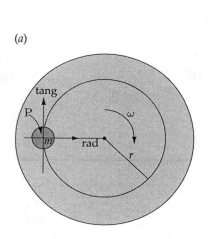

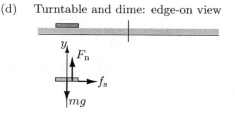

(d)   Turntable and dime: edge-on view

Solve:

Find an *algebraic* expression for the magnitude of the total acceleration, which is all in the horizontal plane. Substitute $\omega = \alpha t$ for constant angular acceleration, so you can eventually solve for time.	
Using the free-body diagram from the figure, apply Newton's second law in the vertical direction and solve for the normal force.	

Now that you have the normal force, find the static frictional force.	
Apply Newton's second law in the horizontal plane. The net force is the frictional force. The dime will not begin to slip until the static frictional force has reached its maximum value, so we use the maximum value for f_s. Use the acceleration from the first step and solve for time.	$t = 3.78$ s

Check: The units are correct. This is a reasonable, measurable time.

Taking It Further: Once the dime starts to slip, in what direction will it slide? Why?

9.2   Rotational Kinetic Energy

In a Nutshell

The **moment of inertia** I of an object is the sum of the product of the mass of individual particles and their distances from the rotation axis squared: $I = \sum m_i r_i^2$. The **moment of inertia** is a measure of an object's resistance to changes in its angular velocity about a particular axis.

The kinetic energy of a rotating object is given by $K = \frac{1}{2}I\omega^2$.

Physical Quantities and Their Units

Moment of inertia I dimensions of $M \cdot L^2$ and units of kg·m^2

Fundamental Equations

Moment of inertia $I = \sum_i m_i r_i^2$

Important Derived Results

Kinetic energy of a rotating object $K = \frac{1}{2}I\omega^2$

Try It Yourself #3

The mass of a dime is 2.268 g. Assuming the turntable of problem #2 has a moment of inertia $I = 0.0234$ kg·m^2, and that all the mass of the dime is located at point P, what is the minimum amount of work that must be done to cause the dime to slip?

Picture: Use the time and the angular acceleration given in problem #2 to determine both the linear and angular speed when the dime starts to slip. Use the expressions for kinetic energy to determine the kinetic energy of both the dime and the turntable at that time. The work done is equal to the increase in kinetic energy of the turntable–dime system.

Solve:

Use the constant angular acceleration expressions to determine the angular speed of the turntable when the dime slips.	
Relate the angular speed to the tangential speed of the dime at its location.	
Determine the kinetic energy of the turntable using its moment of inertia and angular speed.	
Determine the kinetic energy of the dime, using its mass and tangential speed.	
Find the total kinetic energy of the system. This is equal to the minimum work that must be done.	$K = W = 0.241$ J

Check: The units are correct, and this is a reasonable amount of work.

Taking It Further: If the dime is moved closer to the center, will more or less work be required?

9.3   Calculating the Moment of Inertia

In a Nutshell

The **moment of inertia** I is a measure of an object's inertial resistance to changes in angular velocity about a particular axis. It depends on the mass and the distribution of the mass about the axis. For a system of discrete particles, the moment of inertia can be calculated as in the previous section.

For a continuous object, we imagine the object to consist of a continuum of very small mass elements, dm, and as such the summation in the previous section becomes the integral $I = \int r^2\, dm$, where r is the radial distance from the axis of rotation to mass element dm.

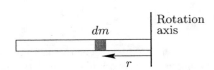

Table 9-1 on page 295 of the text lists moments of inertia for a number of objects about axes through their centers of mass.

It can be shown that the moment of inertia I of an object about a given axis is equal to $I = I_{cm} + Mh^2$ where M is the mass of the object, I_{cm} is its moment of inertia about a parallel axis that passes through the center of mass, and h is the distance between the two axes.

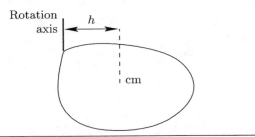

Fundamental Equations

Moment of inertia

$$I = \int r^2\, dm$$

Important Derived Results

Parallel-axis theorem

$$I = I_{cm} + Mh^2$$

Common Pitfalls

> ➤ Two objects with the same mass and the same physical dimensions do not necessarily have the same moment of inertia. The moment of inertia depends on how that mass is distributed.
> ➤ The parallel-axis theorem is often quite useful. But remember that, as its name states, it works only for *parallel* axes.

3. TRUE or FALSE: The moment of inertia of a rotating rigid object about the rotational axis is independent of the rate of rotation.

4. Can there be more than one value for the moment of inertia for a given rigid object? Why or why not?

Try It Yourself #4

A system consists of four point masses connected by rigid rods of negligible mass as shown. (a) Calculate the moment of inertia about the A and y axes by direct application of the formula $I = \sum m_i r_i^2$. (b) Use the parallel-axis theorem to relate the moments of inertia about the A and y axes and solve for the moment of inertia about the y axis. Compare this result with your result in part (a). (c) Calculate the moments of inertia about axes x and z.

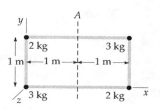

Picture: We will use the definition for moment of inertia and the parallel-axis theorem.

Solve:

(a) Calculate the moment of inertia about the A axis. Each mass is a distance of 1 m from the axis.	$I_A = 10.0$ kg·m^2
(a) Calculate the moment of inertia about the y axis. Two masses lie on the axis, so their distance from this axis is zero, and two masses are 2 m from this axis.	$I_y = 20.0$ kg·m^2
(b) Because the A axis goes through the center of mass of the system, we can use the parallel-axis theorem to calculate I_y from I_A. We better get the same answer as the previous box.	$I_y = 20.0$ kg·m^2
(c) Calculate the moment of inertia about the x axis. Once again, two of the masses lie on the axis, so their distance from this axis is zero. The remaining two masses are each a distance of 1 m from the x axis.	$I_x = 5.00$ kg·m^2

(c) Calculate the moment of inertia about the z axis. Only the 3.00-kg mass at the origin lies on this axis. The 2.00-kg mass on the y axis is 1 m away. The other 2.00-kg mass is 2 m away, and the other 3-kg mass is a distance of $\sqrt{5}$ m away.	$I_z = 25.0$ kg·m^2

Check: Although the moment of inertia could be the same about different axes, in general we expect that the moment of inertia should be different for different axes.

Taking It Further: About which of the axes used in this problem will it be easiest to start this structure spinning, and why?

Try It Yourself #5

Calculate the moment of inertia about the x and y axes for the thin solid triangle of mass M, height a, and width $b = 2a$ as shown.

Picture: To find the moment of inertia about the y axis use vertical strips with a differential thickness dx, where x is the distance from the y axis. To find the moment of inertia about the x axis use horizontal strips.

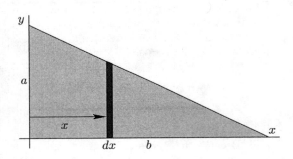

Solve:

Find an equation for $y(x)$ that describes the sloping top edge of the triangle.	
Determine the mass density of the triangle. Because it is made of a uniform material, the density will be the total mass divided by the total area of the triangle.	
Find the differential mass dm of a vertical strip of thickness dx. This will be the differential area of the vertical strip multiplied by the mass density calculated above. The height of this strip should depend on x.	

Find the differential moment of inertia for the small differential rectangular strip and integrate to find the total moment of inertia about the y axis.	$I_y = \frac{1}{6}Mb^2$
Rearrange your equation for $y(x)$ to solve for $x(y)$.	
Find the differential mass dm of a horizontal strip of thickness dy. This will be the differential area of the horizontal strip multiplied by the mass density calculated above. The width of this strip should depend on y.	
Find the differential moment of inertia for the small differential rectangular strip and integrate to find the total moment of inertia about the x axis.	$I_x = \frac{1}{6}Ma^2 = \frac{1}{24}Mb^2$

Check: The mass of the triangle is distributed further out from the y axis than the x axis. Therefore, we expect the moment of inertia about the y axis to be larger.

Taking It Further: If you could choose any rotation axis you wanted, describe qualitatively the axis you would expect to have the smallest moment of inertia, and explain your reasoning.

9.4   Newton's Second Law for Rotation

In a Nutshell

In rotational motion, the dynamic quantity that causes a rotational acceleration is the **torque**. **Torque** is the measure of a force's ability to produce a change in the rotational motion of an object. The torque produced by a force about an axis equals rF_t, where r is the distance from the rotation axis to the point of application of the force and F_t is the tangential component of the force.

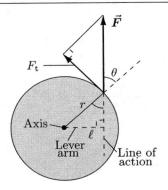

The **line of action** of the force is a line parallel to the applied force, and through the point of application of that force.

The **moment arm** or **lever arm** of a force is the shortest distance from the rotation axis to the line of action of the force.

In this chapter we consider torque about a specific axis, so we can treat it as a one-dimensional quantity. However, torque is a true vector quantity, and we provide a vector treatment of torque in Chapter 10.

Newton's second law for rotation states that the net external torque on a rigid object is equal to the moment of inertia of the object times its angular acceleration.

The torque on a rigid object due to gravity is equal to the weight of the object multiplied by the distance from the rotation axis to the center of gravity of the object.

For any object in a uniform gravitational field, the center of gravity and the center of mass coincide.

Physical Quantities and Their Units

Torque τ dimensions of Force $\times$ length and units of N·m

Fundamental Equations

Torque about an axis $\tau = F_t r = F r \sin\theta = F\ell$

Newton's second law for rotation $\tau_{\text{net ext}} = \sum \tau_{\text{ext}} = I\alpha$

Important Derived Results

Torque due to gravity $\tau_{\text{grav net}} = M g x_{\text{cm}}$

Common Pitfalls

> ➤ Torque does not depend on the distance between the rotation axis and the point of application of the force. Rather, it depends on the length of the moment arm (the perpendicular distance between the rotation axis and the line of action of the force). If the line of action intersects the rotation axis then the torque must be zero.

> Torque, work, and energy all have the same units. However, torque cannot be equated to any form of work or energy. The underlying physical definition of torque has nothing to do with work or energy.

5. TRUE or FALSE: If the net external force acting on an object is zero, the net torque on it must also be zero.

6. A man is hanging onto one side of a Ferris wheel as shown, which swings him down toward the ground. (The Ferris wheel's motor is disengaged so it is free to rotate about its axis.) Can you use constant-angular-acceleration formulas to calculate the time it takes him to reach the bottom? Why or why not?

Try It Yourself #6

A 0.500-kg, 0.250-m-radius disk is made from a uniform piece of thin sheet material. As shown, the disk is glued to a 0.200-kg meter stick, with the disk's center at the 75.0-cm mark. If this pendulum is free to rotate about an axis directed out of the page and located at the zero-cm mark of the meter stick, what is the angular acceleration of the pendulum when it makes an angle $\theta = 25°$ with respect to the vertical?

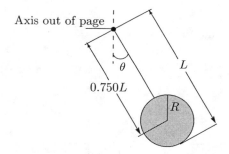

Picture: We need both the moment of inertia and the torque due to the gravitational force to find the angular acceleration. The moment of inertia of the system will be the sum of the moments of inertia of the disk and the meter stick. To find the torque, we need the location of the center of mass of the pendulum.

Solve:

Calculate the moment of inertia of the stick rotated about its end.	

Calculate the moment of inertia of the disk rotated about the end of the stick. The parallel-axis theorem may be helpful.	
Sum the two moments of inertia to get the total moment of inertia.	
Find the location of the center of mass of the stick.	
Find the location of the center of mass of the disk	
Determine the location along the stick of the center of mass of the entire system.	
Determine the horizontal separation of the rotation axis and the center of mass of the system. Use this to determine the torque due to gravity.	
Put it all together and use Newton's second law for rotation to determine the angular acceleration of the pendulum.	
	$\alpha = 5.42$ rad/s^2

Check: The units are appropriate.

Taking It Further: If the disk were to be placed closer to the rotation axis, would the gravitational torque increase or decrease? Why?

Try It Yourself #7

Consider the phonograph turntable of problem #1. Modeling the turntable as a solid disk, what torque is required to stop the turntable at the given rate?

Picture: Torque is determined by the moment of inertia of the disk and the angular acceleration.

Solve:

Determine the moment of inertia of the turntable.	
Using the acceleration you found in problem #1, apply Newton's second law for rotations to find the torque.	$\tau = 2.12 \times 10^{-3}$ N·m

Check: The units work out properly. This seems like a pretty small number. On the other hand, it doesn't take much force to stop a turntable, and the radius is less than a meter.

Taking It Further: Is more or less *force* required to stop the turntable if the force is applied at half the radius rather than at the edge of the turntable? Why?

9.5   Applications of Newton's Second Law for Rotation

In a Nutshell

Applying Newton's Second Law for Rotation

Picture: Angular accelerations for rigid objects can be found by using free-body diagrams and Newton's second law for rotation, which is $\tau_{\text{net ext}} = \sum \tau_{\text{ext}} = I\alpha$. If $\tau_{\text{net ext}}$ is constant, then the constant angular acceleration equations apply. Time intervals and angular positions, velocities, and angular accelerations can then be determined using these equations.

Solve:

1. Draw a free-body diagram with the object shown as a likeness of the object (not just as a dot).
2. Draw each force vector along the line of action of that force.
3. On the diagram indicate the positive direction (clockwise or counterclockwise) for rotations and any linear directions.
4. Apply Newton's second law for rotation to solve for the desired quantity.

Check: Make sure that the signs of your results are consistent with your choice for the positive directions of rotation.

There are many situations in which a rotating object like a pulley rotates without the rope slipping. In these **nonslip** situations, the tangential speed and acceleration of the rope are related to the angular speed and acceleration of the rotating object according to the **nonslip conditions:** $v_t = R\omega$ and $a_t = R\alpha$.

When an applied torque increases the angular speed of a rotating object, it does **rotational work** on that object, given by $dW = \tau \, d\theta$.

The **power** delivered by the torque is given by $P = dW/dt = \tau \, d\theta/dt$.

Fundamental Equations

Work done by a torque $\qquad\qquad\qquad\qquad dW = \tau \, d\theta$

Power input of a torque $\qquad\qquad\qquad\qquad P = \dfrac{dW}{dt} = \tau \dfrac{d\theta}{dt} = \tau\omega$

Important Derived Results

Nonslip conditions $\qquad\qquad\qquad\qquad v_t = R\omega$

$\qquad\qquad\qquad\qquad\qquad\qquad\qquad\quad \alpha_t = R\alpha$

Common Pitfalls

> ➤ When a rope is wrapped around a pulley that has mass, you can no longer assume the tension in the rope is the same throughout the rope. If the pulley undergoes an angular acceleration, a net torque must be applied. Usually this means the tension in the rope on one side of the pulley is larger than the tension in the rope on the other side, even if it is the same rope. This is because the rope needs to accelerate the pulley.

7. TRUE or FALSE: All parts of a rotating rigid object have the same centripetal acceleration.

8. Is it possible for a mass attached to a pulley to have a positive linear acceleration and for the pulley to experience a negative rotational acceleration? Why or why not?

Try It Yourself #8

A 0.400-kg pulley has a 6.00-cm radius and blocks 1 and 2 have masses of 0.800 kg and 1.60 kg, respectively. The pulley wheel is a uniform disk, friction in the pulley axle is negligible, and the string neither slips nor stretches. (a) Find the acceleration of block 2 and the angular acceleration of the pulley. (b) Find the tensions in the string segments on either side of the pulley.

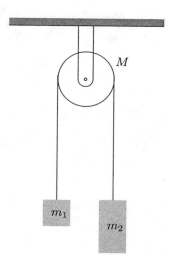

Picture: Draw separate free-body diagrams for each block and the pulley wheel. The tension of the string on either side of the pulley must be different if there is to be a net torque on the pulley. Apply Newton's second law to each block, and Newton's second law for rotation to the pulley. The magnitude of the acceleration of the blocks and the angular acceleration of the pulley are related by the nonslip condition

Solve:

Draw free-body diagrams of all three objects. Because rotations are involved, it is important to draw the forces exactly where they are applied. It is desirable for each object to accelerate in the positive direction at the same time. You can choose the coordinate system for each object so this will be the case. For mass m_1 let "up" be positive. For the pulley, let "clockwise" be positive, and for mass m_2 let "down" be positive. There should be two forces on each of the masses, and four forces on the pulley. However, only two forces on the pulley create a torque.	
Write Newton's second law *algebraically* for each object in the system. Since both masses are attached to the same taut rope, they must undergo the same acceleration. Remember that even though the tension of the rope is different on each side of the pulley, as long as you stay on one side of the pulley the tension is everywhere the same.	
The three equations above have five unknowns, so more equations are needed. The linear and angular acceleration are geometrically related because the string does not slip on the pulley. Write that *algebraic* relationship.	

The pulley is a solid, uniform disk, so we know the moment of inertia. Write that *algebraically*.	
Algebra: Write the pulley equation in terms of a, not α. Also, add the m_1 and m_2 equations. Add these two resulting equations, eliminating both tensions.	
Solve the resulting equation for the acceleration, substituting the moment of inertia. Finally, substitute some numbers to find the linear acceleration.	$a = 3.02$ m/s^2
Now solve for the angular acceleration using the nonslip condition.	$\alpha = 50.3$ rad/s
Finally we can solve for the tensions using our original Newton's second law equations for m_1 and m_2.	$T_1 = 10.3$ N, $T_2 = 10.9$ N

Check: Does Newton's second law for rotation hold for the pulley wheel? If not, then our tensions are incorrect.

Taking It Further: T_1 has to accelerate m_1 upward, in opposition to the force of gravity, while the force of gravity can accelerate m_2 downward. So why is T_2 larger than T_1?

Try It Yourself #9

In the figure the pulley wheel has a moment of inertia of 1.00×10^{-3} kg·m² and radii of 4.00 and 8.00 cm. Blocks 1 and 2 have masses of 1.60 and 1.00 kg, respectively. Assume that friction in the pulley axle is negligible and the strings neither slip nor stretch. (a) Find the acceleration of each block and the angular acceleration of the pulley. (b) Find the tension in each string.

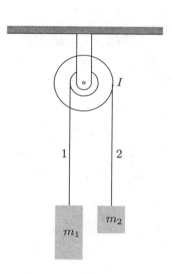

Picture: Draw a separate free-body diagram for each block and the pulley. Apply Newton's second law to each object. Relate the linear and angular accelerations. The accelerations of the blocks are not equal, so there are two relationships to write. Do the algebra.

Solve:

Draw free-body diagrams of all three objects. Because rotations are involved, it is important to draw the forces exactly where they are applied. It is desirable for each object to accelerate in the positive direction at the same time. For m_1, let "down" be positive. For the pulley let "counterclockwise" be positive, and for m_2 let "up" be positive.	

Apply the appropriate Newton's second law to each object in the system *algebraically*. The masses do *not* have the same acceleration in this problem because they are not attached to the same rope. The forces acting on the pulley also have different lever arms.	
Relate the linear and angular accelerations with the nonslip condition.	
Algebra. Use another sheet of paper to solve these three Newton's second law equations simultaneously.	$\alpha = 15.8$ rad/s^2 clockwise $a_1 = 0.630$ m/s^2 up $a_2 = 1.26$ m/s^2 down $T_1 = 16.7$ N ; $T_2 = 8.55$ N

Check: Are the tensions consistent with the angular and linear accelerations? Substitute them into the Newton's second law equations to make sure.

9.6 Rolling Objects

In a Nutshell

When an object rolls without slipping, a point a distance r from the rotation axis moves with a speed $v = r\omega$, where ω is the angular speed of the object. In addition, the speed of the center of mass is given by $v_{\text{cm}} = R\omega$, where R is the radius of the rolling object.

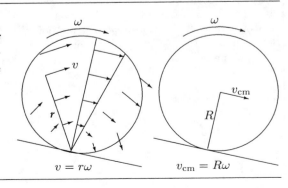

When an object rolls without slipping, a point a distance r from the rotation axis accelerates with $a_t = r\alpha$, where α is the angular acceleration of the object. In addition, the acceleration of the center of mass is given by $a_{cm} = R\alpha$, where R is the radius of the rolling object.

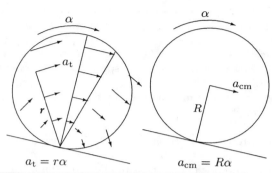

$$a_t = r\alpha \qquad\qquad a_{cm} = R\alpha$$

When an object rolls without slipping, the distance it travels is related to the angle through which it rotates: $s = R\theta$.

The total kinetic energy of a rotating object is equal to the translational kinetic energy of its center of mass plus the rotational kinetic energy about an axis through the center of mass.

Newton's second law for rotation ($\tau = I\alpha$) holds in any inertial reference frame. It also holds in reference frames moving with the center of mass—even when the center of mass is accelerating—as long as the moment of inertia and all torques are computed about an axis through the center of mass.

When an object slips without rolling, $v_{cm} \neq R\omega$. Kinetic friction then tends to change both v_{cm} and ω, increasing one while decreasing the other, until $v_{cm} = R\omega$, at which point rolling without slipping occurs.

Fundamental Equations

Nonslip speed $v = r\omega \; ; \; v_{cm} = R\omega$

Nonslip acceleration $a_{cm} = R\alpha$

Kinetic energy of rotating object $K_{tot} = K_{trans} + K_{rot} = \frac{1}{2}Mv_{cm}^2 + \frac{1}{2}I_{cm}\omega^2$

Common Pitfalls

> The rolling-without-slipping relation $v_{cm} = R\omega$ is valid only for rolling on both flat and curved surfaces, like when cresting a hill.

9. TRUE or FALSE: If a rigid object is rotating, it must be rotating about a fixed axis.

10. Does a rotating object have to roll without slipping? Why or why not?

Try It Yourself #10

A 2.00-kg 6.00-cm-radius wheel has a square hole through its center as shown. The wheel's center of mass is at its geometric center. A string of negligible mass with one end attached to the ceiling is wrapped around its perimeter. As it falls, the string does not slip, so the wheel rotates faster and faster as it gains speed. It is observed that the wheel is moving downward at a rate of 4.74 m/s after falling a distance of 2.00 m. Find its moment of inertia about an axis perpendicular to the page and passing through its center.

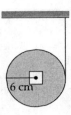

Picture: The string does not slip so no energy is dissipated via friction and the rolling-without-slipping condition holds. Choose a suitable system, apply the conservation of mechanical energy, and solve for the moment of inertia of the wheel.

Solve:

Draw a simple sketch showing the initial and final conditions of the wheel.	
Choose as the system the string, wheel, and Earth. In this system, there are no external forces or energy dissipation from friction, so total mechanical energy is conserved. Write an expression for the conservation of energy in *algebraic* form.	
Write the known quantities. Although we do not know ω_f, we can use the rolling-without-slipping condition to write it in terms of $v_{\mathrm{cm,\,f}}$ and R.	
Substitute the values from the previous step into the conservation of energy equation and solve for the moment of inertia.	$I_{\mathrm{cm}} = 5.37 \times 10^{-3}$ kg·m^2

Check: Make sure your units are consistent and correct throughout the problem.

Try It Yourself #11

A bowling ball (mass 7.00 kg, diameter 20.0 cm) is set spinning about a horizontal axis at 120 RPM. As the ball is set down on the floor, it is still spinning at 120 RPM but its center of mass is at rest. If the coefficient of kinetic friction between ball and floor is 0.800, describe quantitatively the subsequent motion of the ball.

Picture: This is an example of rigid-body motion about a moving axis with a fixed direction. Because a finite force cannot produce an infinite linear acceleration, the initial motion is that of rolling *with* slipping. During the initial rolling-with-slipping stage calculate both the linear acceleration of the center of mass and the angular acceleration about the rotation axis through the center of mass. Using the kinematic formulas, calculate the time when slipping ceases.

Solve:

Draw a free-body diagram of the bowling ball while it is rolling *with* slipping.	
Analyze the linear and rotational motion of the bowling ball. Let rotation in the clockwise direction be positive. This involves writing Newton's second law in *algebraic* form for both the linear and rotational motion of the ball. You should also write kinematic expressions for the linear and angular velocities and accelerations.	
As time increases, the velocity of the center of mass increases and the angular velocity decreases. The slipping will continue until the nonslip condition $v_{\text{cm}} = R\omega$ is met. Set these two quantities equal to each other, using their expressions from the above step. Use this to solve for the time t when rolling-without-slipping begins.	$t = 46$ ms

Substituting the time into the velocity and angular-velocity equations, we can determine the linear and angular speeds of the ball at the time it stops skidding.	
	$v_{cm} = 0.36$ m/s, $\omega = 3.6$ rad/s
Finally, determine the distance the ball traveled and the angle through which it rotated before the skidding stopped.	
	$\Delta x_{cm} = 8.3$ mm, $\Delta \theta = 0.37$ rad

Check: Double-check your units throughout all calculation steps, and in the final answer. Note that it doesn't take the ball long to being to roll without slipping.

QUIZ

1. TRUE or FALSE: Angular velocity and angular acceleration can be defined only in terms of angles measured in radians.

2. TRUE or FALSE: The kinetic energy of a rigid object rotating about a fixed axis depends only on its angular speed and the total mass.

3. If the main rotor blade of the helicopter shown is rotating counterclockwise as seen from above, which way should the smaller tail rotor be pushing the air? Why?

4. A solid ball, a solid disk, and a hoop, all with the same mass and the same radius, are set rolling without slipping up an incline, all with the same initial linear speed. Which goes farthest? If they are all set rolling with the same initial kinetic energy instead, which goes farthest?

5. Defend or refute this statement: The moment of inertia of a rigid object about an axis through its center of mass is smaller than that about any other parallel axis.

6. As shown, a circular disk of mass M_1 and radius R is made from a uniform piece of thin sheet material. This disk is free to rotate about an axis perpendicular to it that passes through its center. A second disk with a radius of $R/2$ is cut from the same sheet and the two disks are glued together with the small disk's perimeter touching the perimeter of the large disk at a single point. Find an expression for the moment of inertia of the composite two-disk object about the given axis.

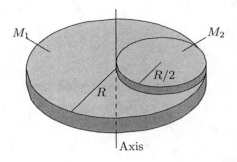

7. A uniform spherical ball is rolling along a horizontal road with speed v_0 when it comes to a hill. It rolls up the hill, just making it to the top, where it comes to rest. How high is the hill? Assume that the ball rolls without slipping and that air resistance is negligible.

Chapter 10

Angular Momentum

10.1   The Vector Nature of Rotation

In a Nutshell

For a rotating object the **angular velocity vector** $\vec{\omega}$ is directed parallel to the rotation axis in the direction specified via the right-hand rule as shown. If you curl the fingers of your right hand in the direction of rotation, $\vec{\omega}$ is in the same direction as your thumb.

The **torque** $\vec{\tau}$ about a point O is mathematically best expressed as the **cross product** (or **vector product**) of $\vec{r}$ and $\vec{F}$. That is, $\vec{\tau} = \vec{r} \times \vec{F}$, where $\vec{r}$ is the vector from the point O to the point of application of force $\vec{F}$.

The cross product of any two vectors $\vec{A}$ and $\vec{B}$ is perpendicular to both vectors. The mathematical convention for determining the direction of a cross product is called the right-hand rule, which is required to completely specify the direction of the cross product. To apply this rule, place the fingers of your right hand in the direction of the first vector ($\vec{A}$) such that when you make a fist they curl toward the direction of the second vector ($\vec{B}$). Your thumb then shows the direction of $\vec{C} = \vec{A} \times \vec{B}$.

If ϕ is the angle between the directions of the two vectors, and $\hat{n}$ is the unit vector in the direction given by the right-hand rule, then $\vec{C} = \vec{A} \times \vec{B} = (AB\sin\phi)\hat{n}$. Consequently $\vec{A} \times \vec{A} = 0$, and $\vec{A} \times \vec{B} = -\vec{B} \times \vec{A}$.

Finding the Vector Product of Two Vectors

Picture: At times it is easier to find a vector product of two vectors by using the equation $\vec{A} \times \vec{B} = AB\sin\phi\,\hat{n}$. At other times it is easier to find the vector product using the Cartesian components of the two vectors.

Solve:

1. The vector product obeys a distributive law under addition: $\vec{A} \times (\vec{B} + \vec{C}) = \vec{A} \times \vec{B} + \vec{A} \times \vec{C}$.

2. If $\vec{A}$ and $\vec{B}$ are functions of some variable such as t, the derivative of $\vec{A} \times \vec{B}$ follows the usual product rule for derivatives:

$$\frac{d}{dt}(\vec{A} \times \vec{B}) = \left(\vec{A} \times \frac{d\vec{B}}{dt}\right) + \left(\frac{d\vec{A}}{dt} \times \vec{B}\right)$$

3. The unit vectors $\hat{\imath}$, $\hat{\jmath}$ and $\hat{k}$, which are mutually perpendicular, have vector products given by $\hat{\imath} \times \hat{\jmath} = \hat{k}$, $\hat{\jmath} \times \hat{k} = \hat{\imath}$ and $\hat{k} \times \hat{\imath} = \hat{\jmath}$. Reversing the order of multiplication gives $\hat{\jmath} \times \hat{\imath} = -\hat{k}$, $\hat{k} \times \hat{\jmath} = -\hat{\imath}$ and $\hat{\imath} \times \hat{k} = -\hat{\jmath}$. Furthermore, $\hat{\imath} \times \hat{\imath} = \hat{\jmath} \times \hat{\jmath} = \hat{k} \times \hat{k} = 0$.

Check: Make sure that your vector products make sense. For example, the vector product of two vectors is a vector and is perpendicular to each of the two vectors. In addition, check your work to make certain you did not inadvertently reverse the order of the two vectors being multiplied, and thus create a sign error.

Taking It Further: Any coordinate system for which these expressions hold is called a right-handed coordinate system. Only right-handed coordinate systems are used in this book.

Fundamental Equations

Torque $\vec{\tau} = \vec{r} \times \vec{F}$

Common Pitfalls

> ≻ When finding a cross product remember that the order of the vectors makes a difference. Use the right-hand rule for cross products to double-check the direction of resulting vector.
> ≻ Thinking of the angular velocity as a vector may seem abstract. However, the direction of the angular velocity vector determines the rotation axis—often an important consideration.

1. TRUE or FALSE: A force directed toward a point exerts zero torque about that point.

2. Is the vector product commutative? Why or why not?

Try It Yourself #1

Consider a small mass located at $\vec{r} = (2.40\hat{\imath} + 3.20\hat{\jmath} + 3.00\hat{k})$ meters. If a force $\vec{F} = (4.50\hat{\imath} + 6.00\hat{\jmath})$ newtons acts on the particle, (a) find the vector torque on the small mass about the origin directly from the components of $\vec{r}$ and $\vec{F}$. (b) Use your part (a) result to verify that $\left|\vec{r} \times \vec{F}\right| = rF\sin\phi$.

Picture: Use the expression for torque and the definition of the dot product.

Solve:

Write an expression for the torque.	
(a) Use the definition of the cross product to find the vector torque about the origin.	$\vec{\tau} = (-18.0\hat{\imath} + 13.5\hat{\jmath})$ N·m
For part (b) first calculate $\left\|\vec{r} \times \vec{F}\right\|$ from your answer above.	$\tau = 22.5$ N·m
Find the magnitude of both $\vec{r}$ and $\vec{F}$.	
Find the angle between $\vec{r}$ and $\vec{F}$ from the two definitions of the dot product.	
Calculate $rF\sin\phi$ and compare it to the magnitude of $\vec{\tau}$. They should be the same.	

Check: If you did not get the same result for the magnitude of the torque using both methods, go back and check your work.

Taking It Further: Is this particle experiencing circular motion about the origin? How can you tell?

Try It Yourself #2

Consider the two vector quantities $\vec{A} = -2.40\hat{i} - 1.60\hat{j} + 2.00\hat{k}$ and $\vec{B} = 4.00\hat{i} + 3.00\hat{j}$. Find the angle between $\vec{A} \times \vec{B}$ and $\vec{C} = 4.00\hat{i} + 3.00\hat{j}$.

Picture: Find the cross product, and then take the dot product of $\vec{A} \times \vec{B}$ and $\vec{C}$ to find the angle.

Solve:

Find the cross product of $\vec{A} \times \vec{B}$ in component form.	
Use the two definitions of the dot product to find the angle between $\vec{A} \times \vec{B}$ and $\vec{C}$.	$\phi = 90°$

Check: This angle is reasonable.

Taking It Further: How could you have arrived at this answer without going through all the mathematics?

10.2 Torque and Angular Momentum

In a Nutshell

The **angular momentum** $\vec{L}$ of a moving particle about the origin is given by $\vec{L} = \vec{r} \times \vec{p}$ where $\vec{r}$ is the particle's position vector and $\vec{p}$ is its linear momentum $m\vec{v}$. Like torque, angular momentum is defined relative to a point in space rather than to an axis (line). The concept of angular momentum can be extended from that of a single particle to that of a system of particles, where the total angular momentum is simply the sum of the angular momenta of the individual particles.

Newton's second law for angular motion states that the net external torque is equal to the time rate of change of the system's angular momentum.

The angular impulse–angular momentum equation says that the change in angular momentum of a system is equal to the time-integrated net external torque acting on the system.

It is often helpful to express the total angular momentum of a system as the sum of two terms: $\vec{L}_{\text{spin}}$, the spin angular momentum associated with any motion relative to the center of mass, and $\vec{L}_{\text{orbit}}$, which is the angular momentum that a point particle of mass M located at the center of mass and moving at the velocity of the center of mass would have.

In Chapter 9, torques are computed about axes instead of about points. To determine the torque about a particular axis, like the z axis, for example, you simply use the z component of the torque about the origin $\vec{\tau}_z$. Similarly, the angular momentum about the y axis is simply the y component of the angular momentum about the origin: $\vec{L}_y$.

The angular momentum of a system rotating about a symmetry axis is simply the product of the moment of inertia about that axis and the system's angular velocity.

Gyroscopic motion illustrates some of the non-intuitive vector properties of rotational motion. A spinning gyroscope moves such that the change in its angular momentum vector is in the same direction as the net external torque vector.

Fundamental Equations

Angular momentum of a point particle	$\vec{L} = \vec{r} \times \vec{p} = \vec{r} \times m\vec{v} = mr^2\vec{\omega} = I\vec{\omega}$
Angular moment of a system	$\vec{L}_{\text{sys}} = \sum_i \vec{L}_i$
Angular moment of a system rotating about a symmetry axis	$\vec{L} = I\vec{\omega}$
Newton's second law for angular motion	$\vec{\tau}_{\text{net ext}} = \dfrac{d\vec{L}_{\text{sys}}}{dt}$
Angular impulse–angular-momentum equation	$\Delta\vec{L}_{\text{sys}} = \displaystyle\int_{t_1}^{t_2} \vec{\tau}_{\text{net ext}}\, dt$
Spin and orbital angular momentum	$\vec{L}_{\text{sys}} = \vec{L}_{\text{orbit}} + \vec{L}_{\text{spin}}$
Orbital angular momentum	$\vec{L}_{\text{orbit}} = \vec{r}_{\text{cm}} \times M\vec{v}_{\text{cm}}$

Important Derived Results

Torque about the z axis $\qquad\qquad\qquad \vec{\tau}_z = \vec{r}_{\mathrm{rad}} \times \vec{F}_{xy}$

Angular momentum about the z axis $\qquad \vec{L}_z = \vec{r}_{\mathrm{rad}} \times \vec{p}_{xy}$

Common Pitfalls

> For an object rotating about a fixed axis, the angular velocity vector and angular momentum vector may be in different directions. The angular momentum vector, unlike the angular velocity vector, depends on the mass distribution. Only for sufficiently symmetric mass distributions are the two in the same direction.

3. TRUE or FALSE: Both angular momentum vectors and torque vectors are defined relative to some particular origin or reference point.

4. The propeller of a light airplane is rotating clockwise as seen by someone observing it from behind the plane. If the pilot pulls back the stick, the flaps on the tail fins are raised. This results in the air pushing down on the plane's tail, thus causing the tail to lower and nose to rise. However, exerting a torque on an object with angular momentum causes it to precess. When the tail flaps are up, which way, port or starboard, will the plane precess? (Port is to the left and starboard is to the right of a person facing the same direction as the plane.) Why?

Try It Yourself #3

A ball is thrown with an initial horizontal velocity and falls under the influence of only gravity. Pick a coordinate system with the origin on the ground and directly below the point where the ball is released. Show explicitly that $\sum \vec{\tau} = d\vec{L}/dt$ holds for this motion about the origin. Let "up" be the $+y$ direction, and the initial velocity of the ball be in the $+x$ direction.

Picture: The only force is the weight, which equals $m\vec{g}$. Draw the position vector from the origin to the particle at some arbitrary time, write expressions for the torque vector $\vec{\tau}$ and the angular momentum vector $\vec{L}$, and take the time derivative of the angular momentum vector.

Solve:

Draw a sketch to visualize the situation.	

Express the position, force, and linear momentum in terms of m, g, v, t, and the unit vectors $\hat{i}$, $\hat{j}$ and $\hat{k}$. Remember that your linear momentum should have both horizontal and vertical components.	
Substitute the expressions of the above step into the equations for torque and angular momentum.	
Take the derivative of the angular momentum with respect to time and show that it is equal to the torque calculated in the step above.	

Check: You should get $\vec{\tau} = -mgvt\hat{k}$.

Taking It Further: Would this result change if a different origin were chosen?

Try It Yourself #4

A 2.00-kg rabbit is sitting on a small, stationary merry-go-round 70.0 cm from the axis of rotation. The merry-go-round has a moment of inertia about the axis of 3.00 kg·m^2. The rabbit becomes startled and executes a horizontal jump in a direction perpendicular with the radial direction (tangentially). This leaves the merry-go-round with an angular speed of 4.00 rad/s. Assuming that air resistance and friction in the merry-go-round's bearings are negligible, and the rabbit's jump required 0.100 s, what was the average force the rabbit exerted on the merry-go-round?

Picture: The change in the merry-go-round's angular momentum is equal to the average torque exerted multiplied by the time duration of that torque. We can get the force of the rabbit's jump from the torque.

Solve:

Draw a sketch showing the initial and final states of the rabbit and merry-go-round.	
Find an *algebraic* expression for the change in the angular momentum of the merry-go-round. Remember it is initially stationary.	
Determine the rate of change of the angular momentum of the merry-go-round.	
Equate this rate of change of the momentum to the torque applied by the rabbit. Since the rabbit jumps tangentially, the angle between $\vec{r}$ and $\vec{F}$ is 90°. Solve for the force exerted by the rabbit.	
	$F = 171$ N

Check: This is roughly equivalent to the weight of a 17-kg rabbit. Most rabbits are less massive than this, but we would expect the jumping force to be larger than the rabbit's weight.

Taking It Further: If the rabbit jumps horizontally in some direction other than tangentially, compare the motion of the merry-go-round to that in this problem.

10.3   Conservation of Angular Momentum

In a Nutshell

The **law of conservation of angular momentum** states that if the net external torque $\vec{\tau}_{\text{ext}}$ acting on a system about some point is zero, then the system's total angular momentum $\vec{L}_{\text{sys}}$ about that point remains constant. This follows from the angular version of Newton's second law. Like the conservation of linear momentum, this law is universally valid.

The net torque due to all internal forces of a system sum to zero.

Important Derived Results

Rotational kinetic energy

$$K = \tfrac{1}{2} I \omega^2 = \frac{L^2}{2I}$$

Common Pitfalls

> ➤ Conservation of angular momentum can still be useful even if there is a net torque. For example, the z component of the angular momentum is conserved as long as the z *component* of the net torque remains zero, even if the other components of the net torque are nonzero.

5. TRUE or FALSE: If the total linear momentum of an arbitrary system of particles is conserved, the total angular momentum must also be conserved.

6. A basketball thrown against a closed door causes it to rotate open. The door is stationary before the collision and rotating after it, so during the collision it clearly gains angular momentum. Where did this angular momentum come from? The ball didn't lose any, or did it?

Try It Yourself #5

A merry-go-round consists of a circular platform mounted on a frictionless vertical axis. It has a radius of 2.40 m, a 210-kg·m^2 moment of inertia, and is rotating with an angular velocity of 2.40 rad/s with a 38.0-kg boy standing on it right at its edge. The boy then runs around its perimeter, running at a constant speed relative to the platform. As he comes up to speed the platform's angular speed decreases to 1.91 rad/s. How fast is the boy moving relative to the platform?

Picture: There are no external torques on the boy–merry-go-round system, so its angular momentum remains constant. Find the initial angular momentum of the system. Obtain an expression for the angular momentum of the system when the boy is running and set it equal to the initial angular momentum. Solve for the speed of the boy relative to the ground. Calculate the speed of the edge of the platform relative to the ground. Knowing it and the boy's speed relative to the ground, find his speed relative to the platform.

Solve:

Draw a sketch to help visualize the problem. A top-down view showing the angular velocity of the boy and merry-go-round, in both the initial and final states, will likely be most helpful.	
Find an *algebraic* expression for the initial moment of inertia of the boy–merry-go-round system. Approximate the boy as a point mass.	
Find an *algebraic* expression for the initial angular momentum of the boy–merry-go-round system. Don't forget your subscripts indicating this is for the *initial* configuration.	
Since the geometry of the boy and merry-go-round do not change, their individual moments of inertia remain the same. However, their angular velocities do *not*. Find an *algebraic* expression for the final angular momentum of the boy–merry-go-round system. Don't forget your subscripts indicating this is the for the *final* configuration.	
Equate the initial and final angular momenta to find an *algebraic* expression for the boy's final angular velocity. Relate his final angular velocity to his speed v relative to the ground.	
Substitute numerical values into your expression above and find the boy's speed relative to the ground.	

Calculate the final speed of the edge of the platform relative to the ground.	
Using the speeds from the previous two steps, calculate the speed of the boy relative to the platform.	$v = 2.30$ m/s

Check: The units work out, and this speed is considerably smaller than the roughly 10 m/s of a world-class sprinter, so seems reasonable.

Taking It Further: Is the boy running in the direction of merry-go-round's rotation or against it? How can you tell?

Try It Yourself #6

As shown in the figure, an 80.0-kg man runs up and jumps onto the bottom of a stationary Ferris wheel whose moment of inertia is 1300 kg·m². The radius of the wheel is 2.40 m and friction on its axle is negligible. After the man jumps on, the wheel rotates freely, swinging him up above the ground. With what minimum initial speed v_0 must he run if it swings him over the top?

Picture: The man–Ferris wheel collision is inelastic, so the total mechanical energy of this system is not conserved. During the collision, the angular momentum of the man–Ferris wheel system is conserved. Equate the values of the angular momentum of the system immediately before and immediately following the collision. Following the collision, mechanical energy is conserved, so we can equate the final mechanical energy with the man at the top of the Ferris wheel with the mechanical energy immediately following the collision.

Solve:

Draw three separate sketches of the problem, one just before the man jumps on the Ferris wheel, one just after he jumps on the Ferris wheel, and one showing the man having been swung to the top of the Ferris wheel. Label these three states 0, 1, and 2, respectively, and indicate values of linear and angular speeds in each case. Use a variable representation to start, and don't forget the subscripts telling you which figure they refer to.	
Equate the initial and final angular momenta of the collision, shown in states 0 and 1, respectively. Assume the man is a point mass. Just before the collision, he is located directly under the axle of the Ferris wheel, right at the edge of the wheel. Re-arrange this expression to find an *algebraic* expression for ω_1 of the man–Ferris wheel system just after the collision.	
After the collision, while the wheel rotates, lifting the man, mechanical energy is conserved. In this step, consider the system to consist of the man, the Ferris wheel, and the earth. Let the bottom of the Ferris wheel be the zero point for the gravitational potential energy. Although the center of mass of the Ferris wheel does not change, the man is lifted a height of $2R$ by the time he gets to the top. If the man just makes it around, the kinetic energy of the system is zero when the man is at the top of his ride. Write an *algebraic* expression for conservation of mechanical energy as the system evolves from state 1 to state 2, taking into account all the factors described.	

We know ω_1 from the initial collision. Substitute this expression into the one of the previous step, and solve for v_0.	
	$v_0 = 19.0$ m/s

Check: The units are appropriate

Taking It Further: Can the man force himself up and over the Ferris -wheel? Why or why not?

10.4   Quantization of Angular Momentum⋆

In a Nutshell

One of the strange discoveries in physics that took place during the early twentieth century is that, like energy, the angular momentum of bound systems is quantized with the fundamental unit of angular momentum $\hbar$ (read h-bar), which is Planck's constant h divided by 2π.

Physical Quantities and Their Units

Quantum of angular momentum $\qquad\qquad\qquad\qquad \hbar = \dfrac{h}{2\pi} = 1.05 \times 10^{-34}$ J $\cdot$ s

---

⋆Optional material.

QUIZ

1. TRUE or FALSE: The angular momentum vector is always parallel to the net torque on the object.

2. TRUE or FALSE: A particle of mass m moving in a circle of radius R at a speed v has an angular momentum with a magnitude of mvR about any point.

3. Will a gyroscope supported at its center of mass precess under the influence of gravity? Why or why not?

4. If the axis of rotation passes through the center of mass of a rotating wheel, will it be balanced? Why or why not?

5. Does a gyroscope precess, rather than fall over, because its spin counteracts the force of gravity? Why or why not?

6. Tarzan, the ape man, swings down from a tree branch to kick a bad guy in the head. In your calculations, model Tarzan as a uniform thin rod 1.90 m long with a mass of 91.0 kg. When he starts swinging, his body makes an angle of 40° with the vertical. At what speed are Tarzan's feet moving when they whack the bad guy in the head, which is at the bottom of the swing?

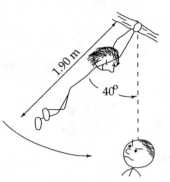

7. Consider the sun as a uniform sphere of radius 1.39×10^9 m and mass 1.99×10^{30} kg rotating on its axis with a period of 26.0 days. If, at the end of the sun's life, all the mass collapses inward to form a neutron star of radius 16.0 km, what will be its period of rotation?

Chapter R

Special Relativity

R.1 The Principle of Relativity and the Constancy of the Speed of Light

In a Nutshell

When one inertial reference frame moves at a speed near that of light compared to another inertial reference frame, interesting things begin to happen to measurements of physical quantities. Clocks in the two reference frames no longer tick at the same rate, the length of objects is not the same in the two frames, and simultaneous events in one reference frame do not appear as simultaneous events in the other reference frame. These phenomena are described by the **special theory of relativity** (often referred to as special relativity), developed by Albert Einstein and others in 1905. This chapter explores many of the consequences of the special theory of relativity.

Postulate I, The principle of relativity: It is impossible to devise an experiment that determines whether you are at rest or moving uniformly—that is, moving at constant velocity relative to an inertial reference frame.

Postulate II: The speed of light is independent of the speed of the light source. The speed of light referred to here is the speed at which light travels through the vacuum of empty space. This postulate explains the Michelson-Morley experiment, which demonstrated that light does not require a medium through which to travel.

One of the consequences of Postulates I and II is the **constancy of the speed of light**: The speed of light c is the same in any inertial reference frame.

R.2 Moving Sticks

In a Nutshell

A stick moving perpendicular to its length has the same length as an identical stick that is stationary. Contrary to intuition, this is not a property of sticks, but rather is a property of space. The "stick" could be replaced by any object, or even simply a region of empty space with a specific length.

The frame of reference in which a stick is at rest is called its **proper frame** or **rest frame**, and the length of a stick in its proper reference frame is called it **proper length** or **rest length**.

R.3   Moving Clocks

In a Nutshell

Clocks moving at high speeds relative to stationary clocks appear to run slow. If a high-speed spaceship travels by us, we would observe that all the clocks on the ship run slower than our clocks. Since clocks measure time, this means that we would observe that time passes more slowly on the spaceship. However, because of the principle of relativity, the people on the ship can consider themselves at rest and us to be moving. From their perspective, our clocks appear to be running slow compared to the clocks on a spaceship. Both these observations are consistent with the constancy of the speed of light and the principle of relativity.

If the time between ticks of a clock in a proper rest frame is T_0 then it can be shown that the time between ticks of an identical clock moving with speed v relative to the rest frame is given by $T = T_0/\sqrt{1 - (v^2/c^2)}$. This phenomenon is known as **time dilation**. In the moving frame of reference, more time passes between ticks of the clock, which means that time (as measured by the number of ticks of the identical clocks) passes more slowly in the moving frame of reference than in the stationary one.

A **spacetime event**, or more simply **event**, is something that occurs at a specific instant in time and at a specific location in space. Any two events that occur both at the same time and same place in one reference frame will also occur at the same time and same place in all reference frames. This is known as the **principle of invariance of coincidences**: If two events occur at the same time and at the same place in one reference frame, then they occur at the same time and at the same place in all reference frames.

Why are coincidences invariant? Because coincidences can have lasting effects. Consider a person accidentally running into a lamppost. This collision requires two spacetime events to occur at the same time and place (a spacetime coincidence). The lamppost needs to be in a particular place at a particular time. The person also needs to be in that same place at the same time. Such a collision will result in the person being bruised after the collision. This bruise will exist regardless of the reference frame used to observe it. Therefore, the coincidence that caused the bruise (both the person and the lamppost being in the same place at the same time) must have occurred in every reference frame as well.

Physical Quantities and Their Units

Speed of light in vacuum
$$c = \frac{1}{\sqrt{\varepsilon_0 \mu_0}} = 3.00 \times 10^8 \text{ m/s}$$

Important Derived Results

Time dilation
$$T = \frac{T_0}{\sqrt{1 - (v^2/c^2)}}$$

Common Pitfalls

> When dealing with problems involving the relationship between space and time measurements made by various observers, it is important to keep in mind the concept of "event." An event could be the location of a lightning bolt striking a point at a given time. The location of each end of a moving rod at the same time constitutes two different events. Each event has four coordinates (x, y, z, t) assigned to it by an observer in reference frame S, and four other coordinates (x', y', z', t') assigned to it by an observer in reference frame S'.

1. TRUE or FALSE: An observer finds that the time interval between two events that take place at the same location is 2 s. Another observer moving relative to this observer will measure the time interval between the two events to be less than 2 s.

2. Suppose that an observer finds that two events A and B occur simultaneously at the same place. Will the two events occur at the same place for all other observers? Explain.

Try It Yourself #1

A rocket car traverses a track 2.40×10^5 m long in 10^{-3} s as measured by an observer next to the track. How much time elapses on a clock in the rocket car when it traverses the track?

Picture: Keep clear in your mind who is measuring each distance and time interval. Because the two clocks are moving with respect to each other, we will need to use time dilation. However, in order to determine the amount of time dilation, we need to know the relative speed of the clocks.

Solve:

Determine, using kinematics, the speed of the car as measured by the observer at the side of the track.	
The clock in the car measures the proper time required to travel that distance. Use the time-dilation expression and the speed calculated above to determine the proper time.	$T_0 = 6.00 \times 10^{-4}$ s

Check: This is quite a rocket car, as humans have not yet been able to achieve these speeds relative to the surface of Earth. The proper time should always be shorter than the dilated time, as we found here. How do we know which is the proper time? The driver of the car measures the start and stop times "in the car," which she perceives as stationary, so the measurements happen at the same place. Therefore the car is the proper frame.

Taking It Further: If the driver of the car measured the time of her drive to be 10^{-3} seconds, what time would be measured by the observers beside the track?

R.4   Moving Sticks Again

In a Nutshell

Consider a stick at rest with a proper length L_0 and an identical stick moving with a speed v in a direction *parallel* to the length of the stick. In this case, an observer in the rest frame will measure the length of the moving stick to be shorter than the length of the stick at rest. This phenomenon is known as **length contraction**. The measured length of the moving stick will be $L = L_0\sqrt{1 - (v^2/c^2)}$. Length contraction occurs only along the direction parallel to the stick. This rule is independent of the material from which the two sticks are made. As a result, the rule is not a rule of sticks, specifically. Instead, it reflects a property of space and time. That is, not only could the "stick" be replaced by any object, it could also be replaced by a region of empty space with a specific length.

Important Derived Results

Length contraction
$$L = L_0\sqrt{1 - (v^2/c^2)}$$

Common Pitfalls

> ➤ Often, problems can be solved using the simple expression distance = velocity × time, provided that the distance, velocity, and time all refer to the same reference frame. For example, suppose you are a muon, moving at a speed $0.998c$ relative to Earth that has to traverse a distance of 9000 m measured by an Earth observer. In your frame of reference, the distance to be traversed is contracted and is only $(9000 \text{ m})\sqrt{1 - 0.998^2} = 600$ m, so the time Δt you would need to traverse this distance is found from $600 \text{ m} = 0.998 \left(3 \times 10^8 \text{ m/s}\right) \Delta t$, giving $\Delta t = 2 \times 10^{-6}$ s.

3. TRUE or FALSE: You have a meter stick located at rest along your y axis. Another person moving along your x axis will measure the length of the meter stick to be less than 1 m.

4. Explain how the measurement of time enters into the determination of the length of an object.

Try It Yourself #2

How far does the driver of the rocket car in Try It Yourself #1 determine she travels in traversing the track?

Picture: Keep clear in your mind who is measuring each distance and time interval. The rocket car will see a contracted length of the track.

Solve:

Determine the contracted length of the track from the velocity found in the previous example.	
	$L = 1.44 \times 10^5$ m

Check: The contracted length should be shorter than the proper length.

Taking It Further: At what speed does the driver of the rocket car think she is traveling with respect to the track?

R.5 Distant Clocks and Simultaneity

In a Nutshell

Let us now consider two identical clocks in a single proper reference frame, at rest relative to each other, and separated by a distance of L_0. We can synchronize these clocks while they are at rest so that they display the same clock reading at the same instant of time in the proper frame. If these same two synchronized clocks now move at a speed v relative to the proper frame, in a direction parallel to L_0, the clock "in front" is actually behind—that is, slower than—the clock "in back." This loss of synchronicity is a direct result of length contraction and the constancy of the speed of light. It is referred to as the **relativity of simultaneity**: If two clocks are synchronized in their rest frame, then in a frame where they move with speed v parallel to the line joining them the clock in the rear is ahead of the clock in the front by a time equal to vL_0/c^2.

Common Pitfalls

➢ Since two events that occur simultaneously for one observer will not in general be simultaneous for other observers, you must make sure you are clear about which observer determines that the two events are simultaneous.

➢ Do not confuse the time at which you see an event take place with the time at which the event actually takes place. For example, suppose you see a distant supernova explode. and record the time on a clock at your location when you make your observation. To determine the actual time that the supernova explosion occurred, you must correct for the travel time that it took the light signal to reach your location.

5. TRUE or FALSE: A clock moving along the x–x' axis at constant speed is struck by a lightning bolt and is later struck by a second lightning bolt. The time interval recorded by this very sturdy clock between the two strikes is the proper time interval.

6. Is it possible for one observer to find that event A happens after event B and another observer to find that event A happens before event B? Explain.

R.6 Relativistic Momentum, Mass, and Energy

In a Nutshell

In special relativity, the conservation laws for energy and momentum still apply, with some slight changes due to motion at high speeds.

The **relativistic momentum** is now given by $p = mv/\sqrt{1 - (v^2/c^2)}$.

The **relativistic kinetic energy** is given by $K = \left(mc^2/\sqrt{1 - (v^2/c^2)} \right) - mc^2$. The **rest energy** of the particle is $E_0 = mc^2$, the energy the particle has when it is at rest.

If we define the total relativistic energy as $E = mc^2/\sqrt{1 - (v^2/c^2)}$, we have $E = K + mc^2$. The total energy squared of a particle may also be written as $E^2 = p^2c^2 + m^2c^4$—that is, the total energy of a particle can be found from its rest mass and its momentum.

Physical Quantities and Their Units

Energy conversion	$1 \text{ eV} = 1.602 \times 10^{-19} \text{ J}$
Rest energy of an electron	$m_\text{e}c^2 = 0.511 \text{ MeV}$
Rest energy of a proton	$m_\text{p}c^2 = 938.3 \text{ MeV}$
Rest energy of a neutron	$m_\text{n}c^2 = 939.6 \text{ MeV}$

Important Derived Results

Relativistic momentum	$p = \dfrac{mv}{\sqrt{1 - (v^2/c^2)}}$
Rest energy	$E_0 = mc^2$
Relativistic kinetic energy	$K = \dfrac{mc^2}{\sqrt{1 - (v^2/c^2)}} - mc^2$
Total relativistic energy	$E = K + mc^2 = \dfrac{mc^2}{\sqrt{1 - (v^2/c^2)}}$
Relativistic energy-momentum relationship	$E^2 = p^2c^2 + m^2c^4$

Common Pitfalls

> In relativistic energy problems do not mistakenly use the classical expressions $K_\text{classical} = \frac{1}{2}mv^2$ or $p = mv$. Whenever v is comparable to c, roughly $v \geq 0.1c$, you must use relativistic expressions.

> Often, particles are described in terms of their kinetic energies rather than their velocities. For example, you may be told that a particle of charge e has been accelerated through a certain voltage, which is equal to the kinetic energy in electron volts that the particle acquires. Relativistic expressions must be used when the kinetic energy is comparable to the rest energy of an object. For example, you must use relativistic expressions for a 2-MeV electron because

an electron's rest mass is about 0.5 MeV. For a 2-MeV proton, however, you can use non-relativistic expressions to a good approximation because a proton's rest mass is about 938 MeV.

7. TRUE or FALSE: An object's kinetic energy is equal to the difference between its total energy and its rest energy.

8. In terms of energies, when can you use $p = E/c$ to a good approximation?

Try It Yourself #3

What is the velocity of a 2.00-MeV electron?

Picture: If $K \ll mc^2$ then you can use the nonrelativistic expression for kinetic energy. Otherwise, you must use the relativistic expression for kinetic energy to determine the electron's velocity.

Solve:

Look up the rest mass of the electron.	
Since the given energy is larger than the rest energy of the electron, you must use the relativistic expression to determine its speed.	$v = 0.967c$

Check: Because the given energy is four times larger than the rest energy of the electron, we expect the velocity to be close to the speed of light, which it is.

Taking It Further: If the particle were a proton instead of an electron, how would the problem change?

Try It Yourself #4

What is the kinetic energy of a proton whose momentum is 500 MeV/c ?

Picture: If $pc \gg mc^2$ then you can use $E = pc$ or $K = pc$ to a good approximation. Otherwise, you will need to use the relativistic expression relating energy and momentum to find the proton's energy.

Solve:

Look up the rest mass of the proton.	
Determine the value of pc.	
Since pc is on the order of the rest energy, the relativistic expression must be used.	
	$K = 125$ MeV

Check: Even for a kinetic energy roughly 1/8 of the rest energy, the relativistic expressions are important. If you had used $K = pc$, you would have found $K = 500$ MeV, which is incorrect by a significant amount, because the rest energy of the proton is significant with these energies.

Taking It Further: How would this problem change if a neutron were used instead of a proton?

QUIZ

1. TRUE or FALSE: An electron and a proton are each accelerated through a potential difference of 50,000 V and are then injected into the magnetic field of a cyclotron. You must use relativistic expressions to analyze the motion of the electron, but the motion of the proton can be treated with classical expressions to a good approximation.

2. TRUE or FALSE: You measure that a rocket ship moving at $0.8c$ takes 5 yr to travel from one galaxy to another. The time measured by the pilot of the rocket ship is 3 yr.

3. An observer stationary in reference frame S$'$ measures a light signal emitted at $t' = 0$ to move from his origin directly along his $+y'$ axis at, of course, a speed c. Describe what an observer in S measures.

4. An observer in reference frame S$'$ is moving relative to reference frame S at $0.8c$. At a certain instant a red flash is emitted at the 10-m mark on the x' axis of S$'$, and 5 s later, as determined by clocks in S$'$, a blue flash is emitted at the same 10-m mark. What is the proper time interval between the red and blue flashes?

5. Suppose that an observer finds that two events A and B occur simultaneously at the same place. Will the two events occur at the same place for all other observers?

6. How fast is a 1 000 000-MeV proton moving?

7. What is the kinetic energy of an electron whose momentum is 300 MeV/c?

Chapter 11

Gravity

11.1 Kepler's Laws

In a Nutshell

Kepler's first law: All planets move in elliptical orbits with the Sun at one focus.

Kepler's second law: A line joining any planet to the Sun sweeps out equal areas in equal times.

Kepler's third law: The square of the period of any planet's orbit is proportional to the cube of the semimajor axis of its orbit.

Physical Quantities and Their Units

Astronomical unit (mean Earth–Sun distance) $1 \text{ AU} = 1.50 \times 10^{11} \text{ m} = 93.0 \times 10^{6} \text{ mi}$

Fundamental Equations

Kepler's third law $$T^2 = Cr^3$$

Common Pitfalls

> In problems where the distances between planets and moons and such appear, the distances are measured between their *centers*.

1. TRUE or FALSE: Kepler's second law, the law of equal areas, implies that the force of gravity varies inversely with the square of the distance.

2. How can a piece of string help you draw an ellipse?

Try It Yourself #1

The average distance of Saturn from the Sun is 1.40×10^9 km. Assuming a circular orbit, find Saturn's orbital speed in meters per second. The average distance of Earth from the Sun is 1.50×10^8 km, and its period is 1.00 year.

Picture: If Saturn's orbit is circular, then its orbital speed is constant. The orbital speed of Saturn is its orbital circumference divided by its period. Determine the distance Saturn travels in one orbital period. Using Saturn's orbital radius and Earth's orbital radius and period, calculate Saturn's orbital period using Kepler's third law.

Solve:

Write an expression for Saturn's speed as a guide for the rest of the problem.	
Determine the distance Saturn travels in one period. This is the perimeter of its orbit.	
Using Kepler's third law and what we know about Earth's orbit, determine Saturn's orbital period.	
Substitute the results from the previous two steps into the expression from the first step to find the orbital speed.	$v = 9780$ m/s

Check: The units work out properly.

Taking It Further: What is the angular momentum of Saturn? You may need to look up the planet's mass.

Try It Yourself #2

The planet Jupiter takes 11.9 years to orbit the Sun. What is the radius of Jupiter's orbit? What is its orbital speed?

Picture: Using Jupiter's orbital period and Earth's radius and period, calculate Jupiter's orbital radius using Kepler's third law.

Solve:

Knowing Earth's orbital period and radius, use Kepler's third law to find the radius of Jupiter's orbit.	$r = 7.82 \times 10^{11}$ m $= 5.21$ AU
Determine the distance Jupiter travels in one period.	
Determine Jupiter's orbital speed from this distance and its period.	$v = 13\ 100$ m/s

Check: The units work out, and we expect Jupiter's orbital speed to be greater than Saturn's.

Taking It Further: Why do we expect Jupiter's orbital speed to be larger than Saturn's? You will probably need to check out the next section on Newton's law of gravity.

11.2   Newton's Law of Gravity

In a Nutshell

Newton's law of gravity states that every object in the universe attracts every other object with a force that is directly proportional to the product of the masses of the two objects and inversely proportional to the square of the distance between them. If $\vec{F}_{12}$ is the force exerted by object 1 on object 2, r_{12} is the magnitude of the relative position vector $\vec{r}_{12}$ from object 1 to object 2, $\hat{r}_{12}$ is the unit vector in the direction of $\vec{r}_{12}$ (that is, $\hat{r}_{12} = \vec{r}_{12}/r_{12}$), and G is the universal gravitational constant, then

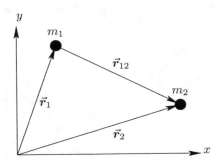

$$\vec{F}_{12} = -\frac{Gm_1m_2}{r_{12}^2}\hat{r}_{12}$$

Newton showed that Kepler's second law, the law of equal areas, is a direct consequence of the conservation of angular momentum and holds for any central force.

Newton was able to prove that the force of gravity exerted by any spherically symmetric object on a point mass on or outside its surface is the same as if all the mass of the object were concentrated at its center.

The universal gravitational constant G can be experimentally determined by measuring the force of gravitational attraction between two objects of known mass. This measurement is difficult because the force is extremely small for objects of laboratory (rather than planetary) size. Knowledge of G is of great scientific value because it enables the experimental determination of the mass of Earth and other objects in the solar system.

The mass of an object is a measure of two distinct properties: (1) that property responsible for the gravitational force the object exerts on other objects, referred to as its **gravitational mass**, and (2) that property responsible for the object's resistance to acceleration, referred to as its **inertial mass**. The equivalence of these two properties is by no means obvious, but it has been confirmed experimentally to a very high precision. Therefore, we do not ordinarily distinguish between the two, and just use the term "mass." This **principle of equivalence** is a very important concept in Einstein's theory of general relativity.

Physical Quantities and Their Units

Universal gravitational constant $G = 6.67 \times 10^{-11} \text{ N} \cdot \text{m}^2/\text{kg}^2$

Radius of Earth $R_\text{E} = 6.371 \times 10^6 \text{ m}$

Mass of Earth $M_\text{E} = 5.97 \times 10^{24} \text{ kg}$

Mass of the Sun $M_\text{S} = 1.99 \times 10^{30} \text{ kg}$

Mean Earth–Sun distance $r_\text{E} = 1.50 \times 10^{11} \text{ m}$

Fundamental Equations

Gravitational force between two objects

$$\vec{F}_{12} = -\frac{Gm_1 m_2}{r_{12}^2}\hat{r}_{12}$$

Important Derived Results

Kepler's third law revisited

$$T^2 = \frac{4\pi}{GM_S}r^3$$

Acceleration due to gravity near Earth's surface

$$g = \frac{GM_E}{R_E^2} = 9.81 \text{ m/s}^2$$

Common Pitfalls

> The force of gravity exerted by a spherically symmetric object (a collection of uniform concentric spherical shells) on an object located outside the spherical object is the same as if the mass of the spherically symmetric object were concentrated at its geometric center. This is true only for objects where the mass is distributed with spherical symmetry.

> In problems where the distances between planets and moons and such appear, the distances are measured between their *centers*. This means you have to add Earth's radius to distances "above the surface" of Earth to properly use Newton's law of gravity.

> An object moving freely in Earth's gravity is not necessarily moving toward Earth! However, such objects, even those in circular orbits, are always accelerating toward Earth. This is true regardless of the direction they are moving.

> When using Newton's law of gravity to calculate the forces between massive objects, be careful of the order of the subscripts and which $\vec{r}$ you use. By convention, the first subscript identifies the object exerting the force, and the second subscript identifies the object that feels the force.

> The gravitational force is *always* attractive.

3. TRUE or FALSE: A confirmation of Newton's law of gravity is the fact that the acceleration of the moon in its orbit is the same as that of objects falling freely near the surface of Earth.

4. Earth's orbit isn't a perfect circle; Earth is a little closer to the sun in January than it is in July. How can you tell this from the apparent motion of the objects in the heavens?

Try It Yourself #3

An astronaut has landed on a 7940-km-diameter planet in another solar system. The astronaut drops a stone from the port of her landing craft and observes that it takes 1.78 s to fall the 4.50-m distance to the planet's surface. Using these data she calculates the mass of the planet. What value does she obtain?

Picture: From the distance the stone falls and the fall time, find the local free-fall acceleration. Use Newton's law of gravity to calculate its mass.

Solve:

Find an *algebraic* expression for the local free-fall acceleration in terms of the time and distance traveled, using kinematics.	
Apply Newton's second law to the stone, using Newton's law of gravity as the applied force. The force and acceleration are in the same direction. Use this fact to find a second *algebraic* expression for the acceleration of the stone.	
Equate the two acceleration expressions and *algebraically* solve for the mass of the planet.	
Finally, substitute values for your variables to find the mass.	$M_{\text{planet}} = 6.71 \times 10^{23} \text{ kg}$

Check: The units are correct, and this mass is reasonable for a small planet, based on comparison with Earth's mass.

Taking It Further: Will you feel heavier or lighter on this planet, compared to Earth? How can you tell?

Try It Yourself #4

Two masses $m_1 = 7890 \text{ kg}$ and $m_2 = 2310 \text{ kg}$ are located at coordinates $(3, 4, 0)$ m and $(18, -12, 0)$ m, respectively. What is the gravitational force of m_2 on m_1?

Picture: We will use Newton's law of gravity. We will find the relative position vector $\vec{r}_{21}$ from the coordinates of the masses.

Solve:

Draw a sketch showing a coordinate system and the locations of the two masses. Label them appropriately. Also draw the vectors $\vec{r}_1$, $\vec{r}_2$, $\vec{r}_{12}$, and $\vec{r}_{21}$. Finally, draw and label a vector in the direction you expect for the force of m_2 on m_1.	
Write the position vector for $\vec{r}_1$ using $\hat{\imath}$, $\hat{\jmath}$, $\hat{k}$ notation.	
Write the position vector for $\vec{r}_2$ using $\hat{\imath}$, $\hat{\jmath}$, $\hat{k}$ notation.	
The problem asks for the force of m_2 on m_1. Find the vector $\vec{r}_{21}$, in $\hat{\imath}$, $\hat{\jmath}$, $\hat{k}$ notation, from the two position vectors above. Remember this vector needs to point *from m_2 toward m_1*.	
Determine the magnitude of the vector $\vec{r}_{21}$.	
Find $\hat{r}_{12}$.	
Substitute all values into Newton's law of gravitation to find the force of m_2 on m_1.	$\vec{F}_{21} = (1.73 \times 10^{-6}\hat{\imath} - 1.84 \times 10^{-6}\hat{\jmath})$ N

Check: This is an attractive force, so the force on m_1 should point toward m_2, generally down and to the right, which our answer does.

Taking It Further: This example might model the gravitational force between two large trucks. Compare the magnitude of this force with that of the weight of a fly ($m_{\text{fly}} \approx 0.4$ g), and explain whether or not it is a good approximation to neglect the gravitational force between most objects we encounter as part of our everyday lives.

11.3   Gravitational Potential Energy

In a Nutshell

The **gravitational potential energy** of a particle at a location is defined as the negative of the work done by the force of gravity on the particle as it moves to that location from a reference location where its gravitational potential energy is zero.

If a particle leaves the surface of Earth with some speed v_0, it gains potential energy and loses kinetic energy. If its initial speed is sufficiently large that the particle still has some kinetic energy when it is infinitely far from the planet, it is said to have *escaped*. The **escape speed** is the minimum initial speed required so that the particle has zero kinetic energy left once it has traveled infinitely far from the planet.

If the potential energy of a particle and Earth is zero when they are infinitely separated, then the potential energy of this system is always negative. If a particle is placed in orbit about Earth, it will also have some kinetic energy. If the total energy of the system is less than zero, then the particle is said to be in a **bound state**. Otherwise, the particle is **unbounded**.

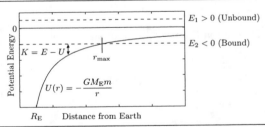

Important Derived Results

Gravitational potential energy ($U \to 0$ as $r \to \infty$) $U(r) = -\dfrac{GMm}{r}$

Escape speed $v_{\text{e}} = \sqrt{\dfrac{2GM_{\text{E}}}{R_{\text{E}}}}$

Common Pitfalls

> Every formula for gravitational potential energy (or for any potential energy, for that matter) assumes a particular zero location. Keep clearly in mind where this location is when working a particular problem. The expression mgh for gravitational potential energy means that the potential energy is zero at $h = 0$.

> Earth's escape speed is 11 km/s or 7 mi/s. Remember that this number refers only to escape from Earth's surface, and that it neglects effects like air drag which increase the escape speed. An object starting from a higher altitude requires a lower initial speed to escape.

> ➢ Every planet, moon, or star has its own escape speed, determined by its mass and radius.
> ➢ Be careful of signs when working problems that deal with the energy of orbital motion. Use the simplest form of the equation for gravitational potential energy, $-GMm/r$, for which the potential energy is always negative and approaches zero as the distance between the object and the earth approaches infinity.
> ➢ For objects in bound orbits, the total mechanical energy is negative, whereas for objects in unbound orbits, the total mechanical energy is *never* negative.
> ➢ For the sake of illustration, we often do problems involving Earth's gravity in which we neglect air drag. Don't worry if the results differ widely from reality. Without air drag, an object falling from a very large distance would strike Earth's surface at a speed approaching Earth's escape speed. However, due to air drag, the actual impact speed is an order of magnitude or so smaller than the escape speed. (The object may also burn up in the atmosphere and never reach Earth's surface.)

5. TRUE or FALSE: Conservation of energy can apply to situations even if the total energy is negative.

6. Is the gravitational potential energy always negative? Why or why not?

Try It Yourself #5

An 1800-kg earth satellite is to be placed in a circular orbit at an altitude of 150 km above Earth's surface. (a) How much work must be done on the satellite to accomplish this? Neglect the resistance of the atmosphere. (b) How much of the work goes into increasing the potential energy of the satellite and how much into increasing its kinetic energy?

Picture: The work that must be done on the satellite is equal to the total change in the Earth–satellite system's mechanical energy because air resistance is negligible. The change in potential energy is due to the satellite's increased height. The change in kinetic energy corresponds to the satellite's orbital speed. This problem is easiest if part (b) is done first.

Solve:

Draw a picture showing the initial and final configurations of the Earth–satellite system. Label them appropriately, with any pertinent, known information.	

Knowing the change in the height of the satellite, find the increase in its potential energy. Don't forget to include Earth's radius!	$\Delta U = 2.59 \times 10^9$ J
Determine the speed required to maintain the circular orbit from what you know about uniform circular motion. Again, don't forget to include the radius of Earth when you have distance calculations.	
Find the increase in the satellite's kinetic energy. Assuming it starts at rest, its initial kinetic energy is zero. Its final kinetic energy is the kinetic energy required to maintain its orbital speed found in the previous step.	$\Delta K = 5.50 \times 10^{10}$ J
The total work done must be equal to the sum of the increases in kinetic and potential energy.	$W = 5.76 \times 10^{10}$ J

Check: Both of these energies may seem large by everyday standards, but they are reasonable for orbital phenomena.

Taking It Further: If you neglected the resulting orbital speed of your satellite when designing the energy provided by your launch vehicle, what would happen to the satellite once it reached the desired orbital height?

Try It Yourself #6

A research satellite is in an elliptical orbit around Earth. Its closest approach to Earth's surface (perigee) is 100 km, and its greatest distance (apogee) is 1600 km. Find its speed at each of these two points.

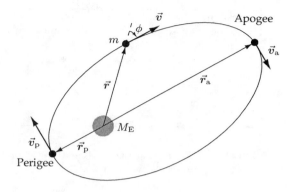

Picture: The only force acting on the satellite is the gravitational force exerted by Earth. This force is directed toward Earth's center, so we know that the torque exerted on the satellite about Earth's center equals zero. It follows that the satellite's angular momentum about Earth's center is conserved. In addition, because the gravitational force is conservative, we know that the mechanical energy of the Earth–satellite system is conserved. Equate expressions for the mechanical energy of the system at apogee and perigee to obtain a second equation. Turn the mathematical crank to find the two speeds.

Solve:

Apply conservation of angular momentum by equating the angular momenta at perigee and apogee. At these points, $\vec{r}$ and $\vec{p}$ are orthogonal. Keep this expression in *algebraic* form.	
Conserve the mechanical energy of the Earth–satellite system by equating the total energy at apogee and perigee. Remember to include both potential and kinetic energy.	
You now have two equations and two unknowns. The rest is algebra, but it is not trivial. Rearrange the angular momentum expression to solve for v_p, and substitute that into the conservation of energy expression to solve for v_a. When calculating r_a and r_p, don't forget to include the radius of Earth. Maintain an algebraic form all through your manipulations until you have an expression for v_a. Then substitute values with their units.	$v_a = 6.69 \times 10^3$ m/s

Substitute the result of the previous step back into the angular momentum expression to solve for v_{p}.	
	$v_{\mathrm{p}} = 8.24 \times 10^3$ m/s

Check: The units are correct, and these speeds are high, but not unreasonable based on other celestial speeds we have seen.

Taking It Further: Does this answer agree or disagree with Kepler's second law, which says the satellite should sweep out equal areas in equal times? Why or why not?

11.4   The Gravitational Field

In a Nutshell

The **gravitational field** at point P is determined by placing a point particle of mass m at P and calculating the gravitational force $\vec{F}_g$ on it due to all other particles. The gravitational force $\vec{F}_g$ divided by the mass m is the gravitational field $\vec{g}$.

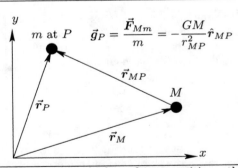

$$\vec{g}_P = \frac{\vec{F}_{Mm}}{m} = -\frac{GM}{r_{MP}^2}\hat{r}_{MP}$$

The point P is called a **field point**. Gravitational fields obey the law of superposition—that is, the gravitational field at a field point due to the masses of a collection of point particles is the vector sum of the fields due to the individual masses.

To find the gravitation field at a field point due to a continuous object, we find the small field $d\vec{g}$ due to a small element of volume with mass dm, and add them over the entire object via integration. dm is found by multiplying the mass density of the object by some small differential size.

$$d\vec{g}_P = -\frac{G\,dm}{r^2}\hat{r}$$

$$\vec{g} = \int d\vec{g}$$

Calculating a Gravitational Field

Picture: Making a sketch of the mass or masses described by a problem is crucial in determining where the field point and source points are. These locations are needed to find both the magnitude and the direction of the gravitational field.

Solve:

1. Draw a diagram that describes the situation given in the problem's statement. Do not forget to identify the field point and the source points. The placement of these points must be accurate because their locations will help you solve the problem.

2. Determine the relative position vector $\vec{r}$ that points *from* the source point *toward* the field point. This vector will be different for each source. You may have to use geometry or trigonometry to determine $\vec{r}$.

3. Use the equation $\vec{g}(r) = -(GM/r^2)\hat{r}$ to determine the full vector gravitational field due to each source.

Check: Do not forget that gravitational fields are vectors, so your answers for these fields must include both their magnitudes and their directions.

Fundamental Equations

Gravitational field of a point source

$$\vec{g} = -\frac{GM}{r^2}\hat{r}$$

Net gravitational field

$$\vec{g}_{\text{net}} = \sum_i \vec{g}_i \rightarrow \int d\vec{g} = \int -\frac{G\,dm}{r^2}\hat{r}$$

Important Derived Results

Gravitational field of a thin uniform spherical shell

$$\vec{g} = 0 \quad \text{for} \quad r < R$$

$$\vec{g} = -\frac{GM}{r^2}\hat{r} \quad \text{for} \quad r > R$$

Common Pitfalls

> Remember that gravitational fields are *vector* quantities. As such, they need to be added vectorally, whether as a discrete sum or within the integral.

> The vector $\vec{r}$ always points *from* the source point (the location of the discrete point mass or the differential mass) *toward* the field point P where the gravitational field is being calculated. Its unit vector is $\hat{r}$, and its length is r.

7. TRUE or FALSE: The gravitational field vector always points toward the mass creating the field.

8. What is needed to create a gravitational field throughout all of space?

Try It Yourself #7

Four masses are positioned at the corners of a square with sides a as shown. Find the gravitational field at the origin.

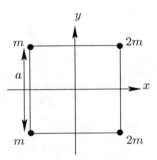

Picture: We will use the expression for the gravitational field of a point mass four separate times, and then add the results to find the net gravitational field.

Solve:

Write a completely general expression for the gravitational field of a point charge to use as a reference for the rest of the problem.	
Determine the $\vec{r}$ for calculating the gravitational field at the origin due to the upper-right mass. Remember that it points away from the mass. Use this to find $\hat{r}$ and r for the field from this mass.	
Use the results of the previous step to find the vector gravitational field from the upper-right point mass.	
Determine the $\vec{r}$ for calculating the gravitational field at the origin due to the upper-left mass. Use this to find $\hat{r}$ and r for the field from this mass.	
Use the results of the previous step to find the vector gravitational field from the upper-left point mass.	

Determine the $\vec{r}$ for calculating the gravitational field at the origin due to the lower-right mass. Use this to find $\hat{r}$ and r for the field from this mass.	
Use the results of the previous step to find the vector gravitational field from the lower-right point mass.	
Determine the $\vec{r}$ for calculating the gravitational field at the origin due to the lower-left mass. Use this to find $\hat{r}$ and r for the field from this mass.	
Use the results of the previous step to find the vector gravitational field from the lower-left point mass.	
Add the vector gravitational fields together to find the net gravitational field at the origin.	$\vec{g} = \dfrac{2\sqrt{2}Gm}{a^2}\hat{\imath}$ N/kg

Check: Since the distribution is symmetric about the x axis, we expect the vertical component of the field to be zero. Also, we expect the field to point toward the more massive objects, which it does.

Taking It Further: Will any components of the gravitational field be zero if we calculate the field at the point $(0, a/2)$? Why or why not?

Try It Yourself #8

A rod with a non-uniform mass density given by $\lambda = 3x$ lies from
$x = 0$ to $x = L$ as shown. Find the gravitational field due to this
rod at the location (L, L).

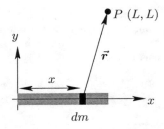

Picture: Because the rod is a distributed object, we will have
to integrate the small differential fields due to differential mass
elements along the rod. The x component of $\vec{r}$ will change for each
differential mass element.

Solve:

Find an expression for the differential mass dm in terms of the mass density and the differential length dx.	
Determine the position vector of the differential mass dm. This should be in terms of the integration variable x. Make sure to include both x and y components, even if they are zero.	
Find $\vec{r}$ for the differential mass element. Remember that it points *from* the mass element *toward* the field point P. Because this $\vec{r}$ will be different for the different dm elements, $\vec{r}$ needs to depend on the integration variable x. Use $\vec{r}$ to find both $\hat{r}$ and r.	
Put it all together to find the small differential gravitational field $d\vec{g}$ due to dm. Integrate to find the total gravitational field at point P. You may need an additional sheet of paper to finish the integral. Remember, you actually have to integrate twice, once for each component of the field.	
	$\vec{g} = (-2.37 \times 10^{-11}\hat{\imath} - 8.29 \times 10^{-11}\hat{\jmath})$ N/kg

Check: We expect the gravitational field to point generally down and to the left, and our solution
does just that.

Taking It Further: If the point P were located at some arbitrary position (x, y), where the values
of x and y are unknown, will we be able to find a numerical result like we did in this case? Why or
why not?

11.5 Finding the Gravitational Field of a Spherical Shell by Integration*

In a Nutshell

If a spherical shell is divided into small rings, its gravitational field can be derived by integrating the gravitational field due to the sum of rings that constitute the shell.

---

*Optional material.

QUIZ

1. TRUE or FALSE: According to Kepler's third law, the period of a planet's orbital motion varies as the 3/2 power of its orbital radius.

2. TRUE or FALSE: The gravitational field is a scalar quantity.

3. What does it mean to say that an astronaut in a satellite orbiting the earth is "weightless"?

4. The drag force of the atmosphere on an orbiting satellite has a tendency to make the satellite's orbit more nearly circular. Why?

5. Some communications satellites remain stationary over one point on Earth. How is this accomplished?

6. If the resistance of the atmosphere is neglected, (a) at what speed would an object have to be projected directly upward from Earth's surface in order to reach an altitude of 200 km? (b) Once it reaches 200 km, how much more energy would an 800-kg object have to be given in order to be put into a circular orbit at this altitude?

7. A satellite is in an elliptical orbit around Earth. At its closest approach to Earth (perigee) it is 112 km above Earth's surface and is moving at a speed of 8032 m/s. When it is at apogee, how far above Earth's surface is it?

Chapter 12

Static Equilibrium and Elasticity

12.1 Conditions for Equilibrium

In a Nutshell

For a rigid body to remain in **static equilibrium** the following conditions must be met:
1. The object must be at rest.
2. The net external force acting on the object must remain zero: $\sum \vec{F} = 0$.
3. The net external torque acting on the object must remain zero: $\sum \vec{\tau} = 0$.

Common Pitfalls

> Equilibrium and static equilibrium are not the same. An object moving with constant nonzero velocity and/or rotating with a constant angular velocity can still have zero external force and torque. In this case the object is not static, because it is moving. However, it is still in equilibrium, because the net external force and torque are zero.
> For an object to be in equilibrium, the net external torque about *any* axis must be zero.

12.2 The Center of Gravity

In a Nutshell

The net torque on an extended object due to gravity about a single point can be calculated as if the entire force of gravity acting on the object were applied at a single point, the **center of gravity**: $\vec{\tau}_{\text{net}} = \vec{r}_{\text{cg}} \times \vec{F}_g$.

In a uniform gravitational field, which applies, for example, to most objects near the surface of Earth, the location of the center of gravity coincides with the location of the center of mass of the object: $\vec{r}_{\text{cm}} = \vec{r}_{\text{cg}}$. So we can write the net torque as $\vec{\tau}_{\text{net}} = \vec{r}_{\text{cm}} \times \vec{F}_g$.

Important Derived Results

Gravitational torque in uniform field
$$\vec{\tau}_{\text{net}} = \vec{r}_{\text{cg}} \times \vec{F}_g = \vec{r}_{\text{cm}} \times \vec{F}_g$$

12.3   Some Examples of Static Equilibrium

In a Nutshell

Choosing the Rotation Axis

Picture: Keep in mind the conditions for static equilibrium: $\sum \vec{F} = 0$ and $\sum \vec{\tau} = 0$.

Solve:

1. To obtain an algebraically simple solution, choose a rotation axis through the line of action of the force you have the least information about.
2. Set the sum of the torques about this axis equal to zero.
3. Set the sum of the forces equal to zero.
4. Solve the set of equations for the unknown quantities.

Check: Try finding alternative ways to solve a problem to check the plausibility of your solution.

A **couple** is a pair of forces that are equal in magnitude, opposite in direction, and are not collinear. In the figure $\vec{F}_1$ and $\vec{F}_2$ constitute a couple. The torque produced by a couple is the same about all points in space. The magnitude of the torque exerted by a couple is $\tau = FD$, where F is the magnitude of one of the forces, and D is the perpendicular distance between the lines of action of the two forces.

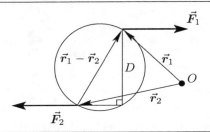

Important Derived Results

Torque of a couple

$$\vec{\tau} = (\vec{r}_1 - \vec{r}_2) \times \vec{F}_1$$

Common Pitfalls

> ➤ Carefully locate your force vectors on free-body diagrams to help ensure accurate calculation of torques. When drawing free-body diagrams, the line of action of a force must pass through the point of application of the force. The effective point of application of the gravitational force acting on an object is the object's center of gravity.

> ➤ For a particular problem, we are free to select the axis about which we calculate torques. Some axis choices are better than others because they result in simpler equations that are more easily solved. A useful rule of thumb is to choose an axis that passes through the point of application of the force that is the least specified. A force is completely specified when its point of application and both its magnitude and direction, or its components, are known.

> ➤ Do not let signs confuse you. To determine whether the sign of a torque $\vec{\tau}$ associated with a force $\vec{F}$ is positive (counterclockwise) or negative (clockwise), imagine that the object is constrained to rotate about the selected axis and that $\vec{F}$ is the only force acting on it. The torque due to $\vec{F}$ is positive if the object would rotate in the positive direction (counterclockwise) and negative if it would rotate in the negative direction (clockwise). An alternative way to determine the sign of the torque $\vec{\tau}$ is to determine the direction of $\vec{r} \times \vec{F}$, where $\vec{r}$ is the vector from the axis to the point of application of the force. (The direction of a cross product is obtained by applying the right-hand rule.) If the direction of $\vec{r} \times \vec{F}$ is up out of the plane of the paper then $\vec{\tau}$ is positive (counterclockwise), and if the direction is into the plane of the paper, then it is negative.

1. TRUE or FALSE: The torque exerted by a couple about a point P is independent of the location of P, unless P is located between the lines of action of the two forces.

2. Must the center of gravity of an object be located inside the material of the object?

Try It Yourself #1

A seesaw consists of a uniform plank, 4.00 m
long, with a mass of 10.0 kg. As shown in
the figure, a 40.0-kg boy sits 0.500 m from the
short end of the seesaw and a girl of mass m_g
sits 0.500 m from the other end. The plank
is balanced when the pivot is located 25.0 cm
from the center of the plank. Determine the
mass of the girl and the force exerted by the
pivot.

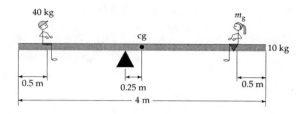

Picture: Draw a free-body diagram of the plank. Because the children are each in equilibrium,
each child pushes on the plank with a force equal to the weight of that child. The weight of the
plank is applied at the plank's center of gravity. Because the plank is uniform, its center of gravity is
at its geometric center. Select an axis that passes through the point of application of the force that
is the least specified. Because we know neither its magnitude nor its direction, the least specified
force is that exerted by the pivot. Because the plank is in equilibrium, both the sum of the forces
and the sum of the torques on the plank are equal to zero.

Solve:

Draw a free-body diagram of the plank. Include the lever arms for each of the forces.	
Determine the torque about the rotation axis due to the boy.	
Determine the torque about the rotation axis due to the plank.	
Determine the torque about the rotation axis due to the girl.	

Determine the torque due about the rotation axis to the pivot support.	
Set the sum of the torques equal to zero and solve for the mass of the girl.	$m_g = 27.1$ kg
Now that you have the mass of the girl, sum the forces to zero to find the force exerted by the pivot.	$F_{pivot} = 756$ N

Check: Your intuition should tell you to expect the mass of the girl to be smaller than the mass of the boy, which it is.

Taking It Further: Why should your intuition lead you in this direction?

Try It Yourself #2

A uniform 20.0-kg strut supporting a 30.0-kg object is suspended by a string and a hinge as shown in the figure. Determine the tension in the string and the force exerted on the strut by the hinge.

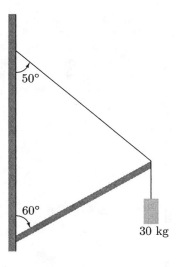

Picture: Draw a free-body diagram of the strut. Select an axis that passes through the point of application of the force that is the least specified. In this case, the least specified force is the force exerted by the hinge. We know neither its magnitude nor its direction. Indicate the lever arm for each force on your diagram. Remember that the lever arm is the perpendicular distance from the axis to the line of action of the force. Because the strut is in equilibrium, the sum of the forces and the sum of the torques are both equal to zero.

Solve:

Draw a free-body diagram of the strut. There are four forces acting on the strut: the force of gravity, the force of the hinge, the force of the upper string, and the force of the lower string. Because the 30-kg mass is in equilibrium, the force of the lower string is equal to the weight of the 30-kg mass. The strut makes an angle of 60° with the vertical axis and thus makes an angle of 30° with the horizontal axis. Because the sum of the angles in a triangle equals 180° we know the angle between the strut and the string supporting the strut is 70°.	
Set the sum of the torques about the hinge equal to zero. Solve for the tension in the support string.	
	$T = 362$ N

Set the sum of the forces equal to zero and solve for the force of the hinge on the strut. Remember that this must be done for the x and y components separately. The tension has both x and y components that can be determined from the free-body diagram.

$$\vec{F}_{\text{hinge}} = (277\hat{\imath} + 258\hat{\jmath}) \text{ N}$$

Check: We should expect the hinge to "push" on the strut, which means exert a force in both the positive x and positive y directions, which it does.

Taking It Further: Is it possible to change the angles of the problem, with the top string still attached to the wall, so that the force exerted by the hinge is zero? How? Or why not?

12.4   Static Equilibrium in an Accelerated Frame

In a Nutshell

The two conditions required for an object to be in static equilibrium in an accelerated reference frame are:

1. $\sum \vec{F} = m\vec{a}_{\text{cm}}$, where $\vec{a}_{\text{cm}}$ is the acceleration of the center of mass, which is also the acceleration of the reference frame.
2. $\sum \vec{\tau}_{\text{cm}} = 0$. The sum of the torques about the center of mass must be zero.

12.5 Stability of Rotational Equilibrium

In a Nutshell

An object is in **stable rotational equilibrium** if small angular displacements of the object away from equilibrium result in a torque (or torques) that tends to return the object to its equilibrium position. For example, the pendulum of a tall clock when it is at rest hanging vertically is in stable rotational equilibrium. If rotated slightly from the vertical and released, the torque due to the gravitational force will tend to rotate the pendulum back to vertical.

An object is in **unstable rotational equilibrium** if small angular displacements of the object away from equilibrium result in torques that tend to rotate the object farther away from the equilibrium position. An example of an object in unstable rotational equilibrium is a pencil balanced on its point. If rotated slightly from the vertical and released, the torque due to the gravitational force will tend to rotate the pencil further from the vertical.

An object is in **neutral rotational equilibrium** if both the net force and the net torque remain zero following small linear and/or angular displacements of the object from an equilibrium position. A meter stick suspended by a nail through a small hole drilled through its center of mass and driven into a wall is an example. The torque due to the gravitational force is zero, so if the stick is rotated slightly from an equilibrium position and released, there is no torque either to return the stick to its initial position or move it farther away.

When you rotate something slightly, if the potential energy increases, remains the same, or decreases, then the rotational equilibrium is stable, neutral, or unstable, respectively.

12.6 Indeterminate Problems

In a Nutshell

The conditions for static equilibrium consist of two vector equations $\sum \vec{F} = 0$ and $\sum \vec{\tau} = 0$. By taking the x, y, and z components of each of these we end up with six component equations. However, with these six equations we can solve for, at most, six unknowns. Any problem that has more unknowns than available independent equations cannot be solved and is said to be indeterminate. However, for static equilibrium problems that concern a deformed object, additional equations can often be obtained by relating the deformations to the forces acting on it. Such equations are introduced in the following section.

12.7 Stress and Strain

In a Nutshell

Most solids behave elastically. If stretched or deformed by external forces, they tend to return to their original size and shape when the deforming stress is removed. For a given material, this is true up to some **elastic limit**; exceeding this limit will cause permanent deformation.

If an applied force tends to stretch an object, the force per unit area perpendicular to the force is called the **tensile stress**.

The resulting fractional change in the object's length is the **tensile strain**.

Up to the proportional limit—a point somewhat lower than the elastic limit for a substance—stress is proportional to strain. The constant of proportionality, which is the ratio of stress to strain, is called **Young's modulus** Y.

The maximum tensile stress a material can withstand is called its **tensile strength**. Young's modulus for both the tensile stress and the tensile strength of a given material may or may not be the same as Young's modulus for the compressive stress and the compressive strength of that material. Table 12-1 on page 410 of the text lists Young's moduli, tensile strength, and compressive strength for several materials.

A sideways deformation resulting from forces that act parallel to the surface of a material is called a **shear strain**. Within the proportional limit of a material, **shear stress**, which is shear force per unit area, is proportional to shear strain. The constant of proportionality is the **shear modulus** M_s. Table 12-2 on page 411 of the text lists shear moduli for some materials.

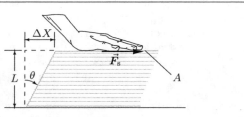

Physical Quantities and Their Units

Strain	dimensionless
Stress	N/m^2
Young's modulus	N/m^2
Shear modulus	N/m^2

Important Derived Results

Tensile stress $$\text{Stress} = \frac{F}{A}$$

Strain $$\text{Strain} = \frac{\Delta L}{L}$$

Young's modulus $$Y = \frac{\text{Stress}}{\text{Strain}} = \frac{F/A}{\Delta L/L}$$

Shear strain $$M_s = \frac{\text{Shear stress}}{\text{Shear strain}} = \frac{F_s/A}{\Delta x/L} = \frac{F_s/A}{\tan\theta}$$

Common Pitfalls

> ➤ Do not think that the elastic moduli are always constants. They are constants only for stresses that are less than the proportional limits of a material. They vary considerably for stresses approaching the elastic limit or the tensile or compressive strength of the material.

3. TRUE or FALSE: Young's modulus and the shear modulus have the same dimensions.

4. Explain why steel is a much better construction material than aluminum or brass.

Try It Yourself #3

A piece of tungsten wire 0.800 m long and a 0.500-m length of steel wire, each of diameter 1.00 mm, are joined end to end, making a total length of 1.30 m. If a mass of 22.0 kg is suspended from the ceiling by this vertical wire, how much does the wire stretch? Neglect the weight of the wire. (Young's modulus is 2.00×10^{11} N/m^2 for steel, 3.6×10^{11} N/m^2 for tungsten.)

Picture: Draw a sketch of the situation. The tensile stress is the same throughout both wires since they have the same diameter and are connected end to end. Obtain expressions for the strain in each wire using the appropriate Young's modulus, and add the extensions for each wire to obtain the net extension.

Solve:

Draw a sketch of the situation.	
The stress and strain are related by Young's modulus. Use this relationship to find an *algebraic* expression for the extension ΔL_t of the tungsten wire. Since the wire has negligible weight, the force on the wire is simply the weight of the mass.	
Using the relationship between stress, strain, and Young's modulus, find an *algebraic* expression for the extension ΔL_s of the steel wire.	
The total extension of the wire will be equal to the sum of the extensions of the two materials. Both have the same force F exerted on them, as well as the same cross-sectional area A. Find the total extension of the wire.	$\Delta L_\text{tot} = 1.30$ mm

Check: The wire is stretched by 0.1% of its length. This seems a reasonable distance.

Taking It Further: Assume that the connection between the two wires is not the weakest point. If the mass is increased which wire will break first, the tungsten or the steel? Why?

Try It Yourself #4

As a runner's foot pushes off the ground, the shearing force acting on an 8.00-mm-thick sole is 25.0 N, distributed over an area of 15.0 cm^2. Find the shear angle θ if the shear modulus of the sole is 1.9×10^5 N/m^2.

Picture: Draw a sketch of the situation. The shear angle can be found from the definition of the shear modulus.

Solve:

Draw a sketch of the situation.	
Use the definition of the shear modulus to find the shear angle.	$\theta = 5°$

Check: This relatively small angle seems reasonable.

Taking It Further: If the runner had smaller feet, would the shear angle increase or decrease? Why?

QUIZ

1. TRUE or FALSE: When an object is in static equilibrium, the net external force acting on it must equal zero.

2. TRUE or FALSE: The lever arm of a force about a point P always equals the distance from P to the point of application of the force.

3. If exactly three external forces act on an object in equilibrium and if two of their lines of action intersect, must the third line of action intersect at the same point?

4. When solving static equilibrium problems, you need to consider the sum of the torques. Is there always a correct rotation point to choose? Explain.

5. You have a job digging holes for posts to support some signs. Explain why the higher above the ground a sign is mounted, the deeper you have to dig the post hole.

6. A nonuniform meter stick is suspended from two light strings attached to the 10.0- and 90.0-cm marks. Masses of 68.4 g and 100 g are suspended from the strings as shown in Figure 12-9. The friction in the pulleys is negligible. Determine both the mass of the meter stick and the location of its center of gravity.

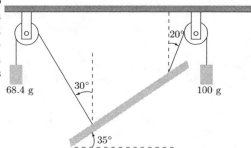

7. A sign is constructed from a uniform 15.0-kg sheet of plywood 80.0 cm high and 120 cm wide. As shown in the figure, the sign is suspended via a pin through its lower left corner and a horizontal cord attached to the upper left corner. Determine the tension in the cord and the force exerted by the pin on the sign.

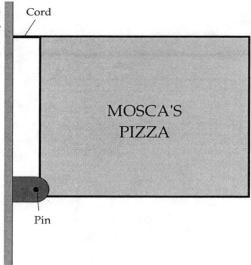

Chapter 13

Fluids

13.1   Density

In a Nutshell

The ratio of the mass of an object to its volume is its **average density**.
One cubic centimeter of water at 4°C has a mass of 1 gram (this was the original definition of the gram), so the density of water is 1 g/cm^3.
An object with a density lower than that of water floats in water; an object of greater density sinks.
The ratio of the density of a substance to that of water is the **specific gravity** of the substance.
Solids and liquids have densities that are roughly independent of external conditions, but the density of a gas is strongly affected by its pressure and temperature. Thus, the temperature and pressure must be specified when stating the density of a gas. Standard conditions are atmospheric pressure at sea level and a temperature of 0°C.

Physical Quantities and Their Units

Density ρ $\qquad\qquad\qquad\qquad\qquad\qquad \dfrac{\text{kg}}{\text{m}^3}$

Density of water $\qquad\qquad\qquad\qquad \rho_{\text{w}} = \dfrac{1 \text{ g}}{\text{cm}^3} = \dfrac{1000 \text{ kg}}{\text{m}^3}$

Volume $\qquad\qquad\qquad\qquad\qquad\quad 1 \text{ L} = 10^3 \text{ cm}^3 = 10^{-3} \text{ m}^3$

Fundamental Equations

Average density $\qquad\qquad\qquad\qquad \dfrac{\text{Mass}}{\text{Volume}}$

Density $\qquad\qquad\qquad\qquad\qquad\quad \rho = \dfrac{dm}{dV}$

Common Pitfalls

> ➤ It is easy to confuse density and specific gravity. Density is the mass per unit volume of a substance. Contrary to the way it may sound, specific gravity is not defined in terms of weight. It is the ratio of the density of a substance to the density of water. It is thus a dimensionless number that is approximately numerically identical to the density expressed in grams per cubic centimeter.

1. TRUE or FALSE: The density of a substance can be expressed as the product of its specific gravity and the density of water.

2. Can an object made of a material with a specific gravity greater than 1 be made to float? Why or why not?

Try It Yourself #1

The density of water is 1.000 g/cm^3 at 4°C. Suppose that a 500-mL flask is filled exactly full of water at a temperature of 60°C, where the density of water is 0.980 g/cm^3. When the flask of water is cooled to 4°C, how much more water must be added to fill the flask again? Neglect the thermal expansion or contraction of the flask itself.

Picture: Determine the mass of the water in the flask at 60°C, and determine the volume of this mass of water at 4°C.

Solve:

Using the known density and volume of the water, calculate its mass at 60°C.	
Determine the volume of this mass of water at 4°C.	
The amount of water to be added is the difference in the two volumes.	$V_{\text{added}} = 10$ mL

Check: This represents a "shrinkage" of the water equal to about 1% of its initial volume, which seems reasonable.

Try It Yourself #2

The mass of a small analytic flask is 15.2 g. When the flask is filled with water, the mass of the flask and water total 119.0 g; when it is filled with an unknown fluid, the total mass is 96.7 g. What is the specific gravity of the unknown fluid?

Picture: You can determine the mass of the unknown volume of water and the unknown fluid. This allows you to use the definition of specific gravity.

Solve:

Determine the mass of the water.	
Determine the mass of the unknown liquid.	
Since equal volumes of liquid are involved, the ratio of the mass of the unknown fluid to that of water will equal the specific gravity.	Specific gravity = 0.785

Check: Since the same volume of the unknown fluid had less mass than water, its specific gravity should be less than one.

13.2 Pressure in a Fluid

In a Nutshell

When a fluid such as water is in contact with a solid surface, the fluid exerts a force normal to the surface at each point on the surface. The force per unit area is called the **pressure** of the fluid.

An increase in the pressure on a material tends to compress it in all directions at once. If the pressure on an object changes, the ratio of the change in pressure ΔP to the fractional decrease in volume $(-\Delta V/V)$ is called the **bulk modulus**.

For liquids whose densities are approximately constant throughout the fluid, the pressure exerted by the fluid increases linearly with the height of the water above the point where the pressure is measured.

Pascal's principle states that a pressure change applied to a *confined* fluid is transmitted undiminished to every point in that fluid and to the walls of the container.

Many pressure gauges measure the difference between the "absolute" pressure P and the atmospheric pressure P_{at}. The difference between absolute pressure and atmospheric pressure is called the gauge pressure P_{gauge}.

Physical Quantities and Their Units

Pressure

1 pascal (Pa) $= 1 \mathrm{~N/m}^2$

1 atmosphere (atm) $= 760$ mmHg $= 760$ torr $= 1.013 \times 10^5$ Pa $= 14.7$ lb/in.^{2}

1 bar $= 10^3$ millibars $= 100$ kPa

Fundamental Equations

Pressure $P = F/A$

Important Derived Results

Bulk modulus $B = -\dfrac{\Delta P}{\Delta V/V}$

Pressure in a static liquid $P = P_0 + \rho g h$

Gauge pressure $P_{\text{gauge}} = P - P_{\text{at}}$

Common Pitfalls

➤ Remember that the decrease of pressure with height (or the increase with depth) is linear only for a fluid that is incompressible and therefore has a constant density. The decrease of atmospheric pressure with height is not at all linear. This is because air is compressible, so its density decreases with increasing height.

➤ When solving problems that deal with the pressure at some depth in a fluid, remember that $\rho g h$ is the pressure difference over a vertical distance h. It is not necessarily the pressure at either the top or bottom.

3. TRUE or FALSE: The pressure in a static fluid is the same at every point throughout the fluid.

4. The following demonstration is done in a science show: Some red-dyed water is poured into a glass U-tube that is open at both ends to show that the fluid "seeks its own level" (that is, it rises to the same height on both sides of the U-tube). Some more red liquid is poured into the U-tube, and the fluid in the U-tube comes to rest as shown. What's the trick here?

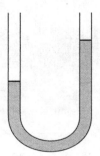

Try It Yourself #3

A lead brick measuring $5.00 \times 10.0 \times 20.0$ cm is dropped into a swimming pool 3.5 m deep. By how much does the volume of the brick change because of the pressure of the water?

Picture: The pressure at the bottom of the pool will cause the volume of the brick to shrink. Use Table 13-2 on page 426 of the text to find the bulk modulus.

Solve:

Find an *algebraic* expression for the difference in pressure on the brick at the bottom of the pool compared to atmospheric pressure. The change in pressure is due entirely to the depth of water in the pool.	

Using the definition of bulk modulus, determine the change in volume of the brick. The volume is calculated by multiplying the three dimensions of the brick.	
	$\Delta V = 4.46$ mm^3

Check: You shouldn't expect a lead brick to be very compressible, so this fractional change of 4.46 ten-thousandths of a percent seems reasonable.

Taking It Further: Would the same size block of steel change its volume more or less when dropped in the pool? Why?

Try It Yourself #4

Blood pressure is normally measured on a patient's arm at approximately the level of the heart. If it were measured instead on the leg of a standing patient, how significantly would the reading be affected? Normal blood pressure is in the range of 70 to 140 torr, and the specific gravity of normal blood is 1.06.

Picture: The question is really simply asking: how much does pressure change with height in blood? You will need to estimate the height difference between arm and leg. Because the blood is flowing and is viscous, your answer won't be exact, but it will give you an idea of how significant the difference is.

Solve:

Write an equation for the pressure difference, as a guide.	
Calculate the density of blood using its specific gravity.	

Calculate the change in pressure, watching your units, and converting to torr. The resulting increase in pressure is significant.	$\Delta P \approx 70$ torr, more or less, depending on your estimate of the height difference.

Check: We should expect the pressure to be higher at the leg than at the arm.

Taking It Further: Will your blood pressure vary significantly if it is measured at the same place on your arm while you are standing up and then while you are sitting on the floor? Why or why not?

13.3   Buoyancy and Archimedes' Principle

In a Nutshell

The force exerted by a fluid on an object completely or partially submerged in that fluid is called the **buoyant force**.

Archimedes' principle states that an object completely or partially submerged in a fluid is buoyed up by a force equal to the weight of the displaced fluid.

The apparent weight of an object submerged in a fluid is the difference between its weight and the magnitude of the buoyant force.

Solving Problems Using Archimedes' Principle

Picture: Carefully read the problem statement to determine the situation. Sketching a picture of the situation is often helpful.

Solve:

1. Apply Archimedes' principle to relate the buoyant force to the weight of the displaced fluid.
2. Apply Newton's second law to the object and solve for the desired quantity.

Check: Verify that your answer is plausible.

Common Pitfalls

> ➢ The equilibrium condition for an object floating in a fluid is that the weight of the object is equal to the weight of the volume of fluid that the submerged portion of the object displaces. The volume of the fluid displaced is, of course, less than or equal to the volume of the object.

> ➢ Just because an object sinks in a liquid, don't think that there is no buoyant force acting on it. The buoyant force is simply less than the object's weight.

5. TRUE or FALSE: The buoyant force on an object completely submerged in a fluid depends on the density of the fluid and the volume of the object but not on the shape or composition of the object.

6. An ice cube floats in a glass of water. As the ice melts, does the water level rise or fall? Explain.

Try It Yourself #5

Some colleagues once had to carry out an experiment on a lake. To do this they built a raft out of Styrofoam, 1 ft (0.305 m) thick and 8 ft (2.44 m) square. Take the specific gravity of Styrofoam to be 0.035. (a) How deep in the water did the unloaded raft float? (b) With three men (whose average mass was 88 kg each) and 120 kg of experimental equipment on the raft, how deep did it float?

Picture: When the raft is floating in equilibrium, the total weight of the raft and load will be equal to the weight of the water displaced.

Solve:

Find an *algebraic* expression for the weight of the raft from the specific gravity of Styrofoam and the volume of the raft.	
Find an *algebraic* expression for the weight of the water displaced by the raft. Make sure the variable you use for the height of the displaced water is different from the variable you used for the thickness of the raft in the previous step.	
Assuming the raft floats, according to Archimedes' principle the weight of the raft must equal the weight of the displaced water. Use this fact to solve for the height of the displaced water.	$h = 1.07$ cm
Now find an *algebraic* expression for the weight of the raft, researchers, and equipment.	

Equate the loaded weight of the raft to the weight of the displaced water to find the new height of displaced water.	
	$h = 7.52$ cm

Check: Since no mention was made of the raft sinking, we expect that it probably floated, even when fully loaded. Our answers demonstrate that even the fully loaded raft was in no danger of sinking.

Taking It Further: If the raft were used to survey the results of an oil spill, would it still float? Why or why not? The specific gravity of oil is 0.68.

Try It Yourself #6

In the figure, oil with a specific gravity of 0.68 is floating on water in a container. A wooden object is floating at the fluid boundary such that one-third of its volume lies below the boundary. What is the density of the wood?

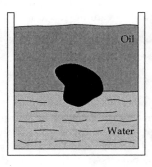

Picture: The wooden object will float such that its weight will equal the total weight of the fluid it displaces. One-third of the fluid by volume is water, and the remaining two-thirds is oil. Since the weights are proportional to g, the total mass of the displaced fluid must equal the mass of the of the wooden object.

Solve:

Write an *algebraic* expression for the total mass of the displaced fluid, which will also be the mass of the wood. This expression should be in terms of the volume of the block, the density of water, and the specific gravity of the oil.	
The mass density of the wood is its total mass, which you found in the previous step, divided by its volume. Substitute numerical values to find the mass density of the wood.	
	$\rho_{\text{wood}} = 787$ kg/m^3

Check: This may seem a bit large, but we don't often encounter a cubic meter of wood. If instead you think of a 10-cm cube of wood, its mass would be .787 kg, which seems reasonable.

Taking It Further: If the oil were removed, would the wood float higher or lower in the water? Why?

13.4 Fluids in Motion

In a Nutshell

The general flow of a fluid can be very complicated, so we restrict our study here to steady, nonturbulent flow. If the fluid is incompressible (a liquid), then the volume flow rate I_V (volume per unit time) and mass flow rate I_M (mass per unit time) in a tube or pipe are constant throughout the length of the tube or pipe.

The **Bernoulli equation** describes the steady, nonviscous flow of an incompressible fluid.

A pressure gradient is required to accelerate a nonviscous fluid; the acceleration is in the direction of decreasing pressure. Thus, in accordance with Newton's second law, a fluid speeds up when it flows into a region of lower pressure and slows down when it flows into a higher pressure region.

As air or another fluid passes through a constriction, its speed increases and its pressure drops. This is known as the **venturi effect**.

When real fluids flow, an internal shear stress is produced that opposes the flow. This shear stress increases with the speed of the flow. This property of fluid flow is known as **viscosity**, and the **coefficient of viscosity** η is a measure of a fluid's resistance to flow. Because of viscosity, a pressure gradient is required to cause a fluid to flow through a horizontal pipe at constant speed. The pressure drop, which occurs along the direction of flow, is proportional to the flow rate. **Poiseuille's law** describes the resistance of a fluid to flowing down a circular tube of radius r.

The **Reynolds number** N_R is a dimensionless parameter that characterizes the degree of turbulence of the flow of a fluid. The transition from laminar to turbulent flow occurs as the Reynolds number increases from 2000 to 3000.

Physical Quantities and Their Units

Mass flow rate I_M	dimensions of M/T
Volume flow rate I_v	dimensions of L^3/T
Viscosity	dimensions of pressure $\cdot$ time ; 1 Pa $\cdot$ s = 10 poise

Important Derived Results

Continuity equation	$I_{M1} - I_{M2} = \dfrac{dm_{12}}{dt}$
Bernoulli equation	$P_2 + \rho g h_2 + \frac{1}{2}\rho v_2^2 = P_1 + \rho g h_1 + \frac{1}{2}\rho v_1^2$
Poiseuille's law	$\Delta P = \dfrac{8\eta L}{\pi r^4} I_V$
Reynolds number	$N_R = \dfrac{2r\rho v}{\eta}$

Common Pitfalls

> ➤ Bernoulli's equation and the continuity equation ($Av = $ constant) are applicable only for the flow of incompressible fluids, so they are of limited use for studying gas flow. The Bernoulli equation applies only for nonviscous laminar flow and Poiseuille's law applies only for laminar flow.

> ➤ You can really mess up calculations with Bernoulli's equation if you are not careful to use consistent units for all the quantities. It is best to put everything into SI units.

> ➤ When doing numerical problems, be sure to write all units and determine the units of your answer. Use the conversions $1\ \text{Pa} = 1\ \text{N/m}^2$, $10\ \text{poise} = 1\ \text{Pa}\cdot\text{s} = 1\ \text{N}\cdot\text{s/m}^2$ and the like.

7. TRUE or FALSE: Shearing forces cannot exist in a fluid.

8. The fluid pressure in a pipe decreases as you go downstream, even when the pipe is level and the fluid is incompressible. Why?

Try It Yourself #7

Water is flowing smoothly at 15 ft/s in a horizontal pipe with a 2-in. inside diameter at an absolute pressure of 40 lb/in.². Neglect the viscosity of water. At a certain point the diameter of the pipe necks down to 1.05 in. (a) How fast does the water flow in the narrow section? (b) What is its pressure there (in lb/in.²)?

Picture: First, convert all quantities to SI units to avoid future complications. The continuity equation can be used to find the water's speed in the narrow section. Bernoulli's equation can be used to find the pressure difference to cause this change in speed.

Solve:

Convert all quantities to SI units.	

Use the continuity equation to find the water's speed at the narrow section.	
	$v = 16.7$ m/s
Use Bernoulli's equation to find the new pressure in the narrow section.	
	$P = 21.3$ lb/in.^{2}

Check: We expect the speed to increase, and the pressure to decrease, at the narrow portion of the pipe, and it appears to do just that.

Taking It Further: We assumed laminar flow in the above calculations. Is this reasonable? How can you tell?

Try It Yourself #8

Water is supplied to an outlet from a pumping station 5.00 km away. From the pumping station to the outlet there is a net vertical rise of 19 m. Take the coefficient of viscosity of water to be 0.0100 poise. The pipe leading from the pumping station to the outlet is 1.00 cm in diameter, and the gauge pressure in the pipe at the point where it exits the pumping station is 520 kPa. At what volume flow rate does water flow from the outlet?

Picture: Bernoulli's equation will not work because the water is viscous. There will be a pressure drop due to viscous flow in the pipe and a drop due to the gain in height. You know the total pressure drop from the pumping station to the outlet and can calculate the drop due to the vertical rise. The difference will be due to the viscous drag in the pipe. Assume the flow is laminar and use Poiseuille's law to find the volume flow rate at the outlet.

Solve:

Find the pressure drop due to the vertical rise.	
Find the pressure drop due to the viscosity. The water exits at atmospheric pressure.	
Use Poiseuille's law to determine the volume flow rate.	$I_V = 1.64 \times 10^{-5}$ m^3/s

Check: Picture a cylinder 1.00 cm in diameter. How long would that cylinder have to be to contain 1 second's worth of water? Does this seem reasonable given your experience with hoses?

Taking It Further: According to Table 13-3 on page 446 of the text, the viscosity of water varies with temperature. If you are the fire department depending on a certain volume of water delivery to put out a fire, would you rather fight a fire in the winter or the summer? Why?

QUIZ

1. TRUE or FALSE: Bernoulli's equation is derived by requiring that the total mechanical energy of an element of the fluid remains constant.

2. TRUE or FALSE: A body that floats in a liquid does so at a depth at which it displaces an amount of fluid equal in weight to its own weight.

3. Assuming that you float in fresh water with about 5% of your body above water, estimate the volume of your body by assuming a value for your mass.

4. No pump that works by suction can raise water in a pipe higher than about 34 feet. Why not?

5. You can't breathe underwater by drawing air through a tube stretching from the water's surface (a "snorkel") to a depth of much more than a foot. Why not?

6. The density of air under "ordinary" (not standard) conditions is about 1.2 kg/m^3, whereas that of helium is 0.17 kg/m^3. What must the radius of a spherical helium-filled balloon be if it is to lift a total load of 350 kg plus the helium itself? Assume the mass of the balloon to be negligible.

7. A town's water tank is supported above ground on posts. Its diameter is 18.0 m, and the water in the tank is 7.50 m deep. The top is open to the atmosphere. If a hole is punched in the bottom of the tank and the water flows out in a stream 1.00 cm in diameter, how long does it take for the water level to drop by 10 cm? Neglect the viscosity of the water, and assume the pressure to be constant.

Chapter 14

Oscillations

14.1 Simple Harmonic Motion

In a Nutshell

Simple harmonic motion is the oscillatory motion that occurs under a restoring force that is proportional to the displacement of the system from equilibrium. The force of a spring is a classic example of a linear restoring force. The resulting motion has acceleration that is proportional to the position.

In simple harmonic motion, the acceleration, and thus the net force, are both proportional to, and oppositely directed from, the displacement from the equilibrium position.

The **frequency** f of an oscillating system is the number of cycles completed per unit of time. The **period** T of the motion is the time required for one cycle of the motion. Frequency is the reciprocal of the period, and has units of cycles per second, or **hertz** (Hz).

The frequency and period of simple harmonic motion are independent of the amplitude of the motion.

The position of an object experiencing simple harmonic motion can be described by the function $x = A\cos(\omega t + \delta)$, where A is the **amplitude** of the motion, $\omega = 2\pi f$ is the **angular frequency** of the motions, and δ is the **phase constant**.

For a mass oscillating on a spring, the angular frequency is equal to the square root of the spring constant divided by the mass.

Solving Simple Harmonic Motion Problems

Picture: Choose the origin of the x axis at the equilibrium position. For a spring, choose the $+x$ direction so that x is positive if the spring is extended.

Solve: Do not use the kinematic equations for constant acceleration. Instead, use the equations developed for simple harmonic motion.

Check: Make sure your calculator is in the appropriate mode (degrees or radians) when evaluating trigonometric functions and their arguments.

There is a very close connection between simple harmonic motion and circular motion with constant speed. The projection onto one coordinate axis of a point undergoing uniform circular motion is a simple harmonic motion.

Physical Quantities and Their Units

Frequency f dimensions of 1/T; units: 1 cy/s = 1 Hz

Period T dimensions of T; units: s

Angular frequency ω dimensions of 1/T; units: rad/s

Fundamental Equations

Linear restoring force (Hooke's law) $F_x = -kx$

Frequency $f = \dfrac{1}{T} = \dfrac{\omega}{2\pi}$

Important Derived Results

Position for simple harmonic motion $x = A\cos(\omega t + \delta)$

Velocity for simple harmonic motion $v_x = \dfrac{dx}{dt} = -A\omega\sin(\omega t + \delta)$

Acceleration for simple harmonic motion $a_x = \dfrac{dv}{dt} = \dfrac{d^2x}{dt^2} = -A\omega^2\cos(\omega t + \delta) = -\omega^2 x$

Angular frequency for simple harmonic motion $\omega = \sqrt{\dfrac{k}{m}}$

Common Pitfalls

> ➤ Do not be careless and use any of the constant-acceleration formulas for the motion of a harmonic oscillator! The force is proportional to the displacement from equilibrium, so the acceleration is definitely not constant.
> ➤ In the formulas involving sine and cosine functions, the argument of the trigonometric function (that is, the quantity whose sine or cosine is taken) must be a dimensionless number; if thought of as an angle, it is generally simpler if the angle is in radians.
> ➤ In almost all cases, it is best to put the origin of the x axis at the equilibrium position of the oscillating particle since that is the zero point of the displacement of, and thus of the force on, the oscillating particle.
> ➤ The formulas we write for simple harmonic motion apply to any motion under a restoring force that is directly proportional to the displacement from an equilibrium position. The "spring constant k" used in the formulas is whatever quantity occupies that position in the force equation, whether or not there are springs involved.

1. TRUE or FALSE: Periodic motion is any motion that repeats itself cyclically.

2. A particle is undergoing simple harmonic motion. What distance does it travel in one full period?

Try It Yourself #1

The motion of a mass oscillating on the end of a spring is graphed in the figure. What are (a) the amplitude, (b) the period, and (c) the frequency of its motion? (d) If the mass is 600 g, what is the force constant of the spring?

Picture: (a) and (b) can be read directly from the graph, (c) can be calculated from (b), and the force constant can be calculated from the mass and period.

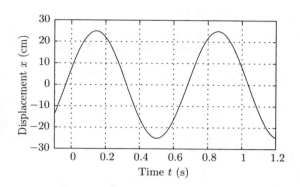

Solve:

The amplitude is the maximum deviation from equilibrium. Find this from the graph.	$A = 25$ cm
The period is the time required for the motion to repeat itself. Read this from the graph.	$T = 0.71$ s
Calculate the frequency from the period.	$f = 1.4$ Hz
Calculate the spring constant from the mass and period.	$k = 46$ N/m

Check: From the graph, the period must be less than 1.2 s.

Taking It Further: Is the phase constant zero? How can you tell?

Try It Yourself #2

You are riding on a 10.0-m diameter Ferris wheel that is rotating at a constant rate of 1.00 rev every 10.0 s. (a) Determine your angular speed in rad/s. Express your height y above the rotation axis as a function of time t so that initially (at $t = 0$) you are (b) 3.00 m above the rotation axis and moving downward, and (c) 2.00 m below the rotation axis and moving upward.

Picture: Draw a sketch to help visualize the situation. Angular speed is total angle divided by total time. Your height y will have the functional form of cosine. The argument of the cosine function will be the angle $\theta = \omega t + \delta$, which is a function of time. Use the conditions given to solve for δ.

Solve:

Draw a sketch to help visualize the situation.	
Find the angular frequency.	$\omega = 0.628$ rad/s
Write a generic *algebraic* expression for the height above the rotation axis as a function of time.	
For part (b), you start 3.00 m above the axis at time $t = 0$. Use this fact to find the phase constant δ. Then substitute all values to find $y(t)$.	$y(t) = (5.00 \text{ m}) \cos\left((0.628 \text{ rad/s})t + 0.927 \text{ rad}\right)$
For part (b), you start 2.00 m below the axis, rotating upward at time $t = 0$. In order to be moving upward, the angle must be greater than 180°. Use this fact to find the phase constant δ. Then substitute all values to find $y(t)$.	$y(t) = (5.00 \text{ m}) \cos\left((0.628 \text{ rad/s})t + 4.30 \text{ rad}\right)$

Check: Substitute $t = 0$ in your expressions and make sure you get the initial positions specified.

Taking It Further: What is the frequency of your motion?

14.2 Energy in Simple Harmonic Motion

In a Nutshell

An object undergoing simple harmonic motion has a constant total energy, but its potential energy and its kinetic energy vary sinusoidally with time. The kinetic energy is a maximum and the potential energy zero, at the equilibrium position. The kinetic energy is zero, and the potential energy a maximum, when the particle is at its maximum deviation (amplitude) from equilibrium.

The total mechanical energy in simple harmonic motion is proportional to the square of the amplitude.

Important Derived Results

Mechanical energy of a harmonic oscillator $E_{\text{total}} = \frac{1}{2}kA^2 = \frac{1}{2}kx^2 + \frac{1}{2}mv^2 = \frac{1}{2}mv_{\text{max}}^2$

Common Pitfalls

> ➤ The potential and kinetic energy of a harmonic oscillator both vary with time; it is only the total energy that is constant. The total energy may be all kinetic, all potential, or anything in between at different points in the cycle.

3. TRUE or FALSE: If a particle is undergoing simple harmonic motion due to the action of a force $F_x = -kx$, its kinetic energy is $\frac{1}{2}kA^2$, where A is the amplitude of its oscillation.

4. A mass-and-spring system is undergoing simple harmonic motion with an amplitude A. If the mass is decreased by half but the amplitude is unchanged, how does the total energy of the oscillation change? How does the total energy change when the original mass is set oscillating with half the initial amplitude?

Try It Yourself #3

An object of mass 1.00 kg oscillates on a spring with an amplitude of 12.0 cm. Its maximum acceleration is 5.00 m/s^2. Find the total energy of the mass–spring system.

Picture: The total energy of the mass–spring system can be found from the spring constant and the amplitude. To find the spring constant, use Newton's second law.

Solve:

Write out an *algebraic* expression for the total energy as a guide for the problem.	
Find an *algebraic* expression for the spring constant from Newton's second law. To find k, we need to know both a and x at the same time. When a is at a maximum, the displacement is also at a maximum, but negative.	
Substituting from step 2 into step 1, you can find the total energy.	$E_{\text{total}} = 0.300 \text{ J}$

Check: The units are correct, and this is a reasonable energy.

Taking It Further: When the mass is located halfway between the equilibrium and maximum displacement locations, what is its kinetic energy?

Try It Yourself #4

A 400-g mass oscillates on the end of a spring with an amplitude of 2.50 cm. Its total energy is 2.00 J. What is the frequency of the oscillation?

Picture: Since you know the total energy and the amplitude, you can find the force constant of the spring. From the force constant and the mass, you can find the frequency of oscillation.

Solve:

Find an *algebraic* expression for the force constant of the spring from the energy of the oscillation.	

From the force constant and the mass, find the frequency of oscillation.	
	$f = 20.1$ Hz

Check: This frequency seems reasonable, although it is starting to get pretty high for a mass on a spring. However, the amplitude must also be taken into account, and this frequency is much more reasonable for a 2.5-cm amplitude than it would be for a 25-cm amplitude.

Taking It Further: Compare the maximum speeds of this mass on a spring and of an identical mass on an identical spring, but oscillating with a 25-cm amplitude.

14.3   Some Oscillating Systems

In a Nutshell

A mass hanging on a vertical spring will extend the spring to a new equilibrium position y_0. All the expressions we have found for a mass on a spring still work, as long as we use the new equilibrium position as $y' = 0$. The only modification we must make is to the total energy of the system, which increases by an amount U_0, which is the potential energy of the new equilibrium position.

For small oscillations, the **period of a simple pendulum** of length L, which has all its mass at the end of the pendulum, is given by $T = 2\pi \sqrt{\frac{L}{g}}$

A **physical pendulum** has its mass distributed throughout the swinging object. The period of the pendulum is independent of its mass, but depends on its moment of inertia I about the rotation axis, the distance D between its rotation axis and the center of mass, and g, the gravitational field strength.

Important Derived Results

Angular frequency of a simple pendulum	$\omega^2 = \dfrac{g}{L}$
Period of a simple pendulum	$T = 2\pi \sqrt{\dfrac{L}{g}}$
Period of a physical pendulum	$T = 2\pi \sqrt{\dfrac{I}{MgD}}$

Common Pitfalls

> ➤ Remember that the formula for the period of a simple pendulum applies *only* to the simple pendulum. If the actual system is something other than a particle on the end of a massless string, then the period will depend on how the mass is distributed in space.
> ➤ The motion of a mass on a spring hanging vertically is simple harmonic motion even though a second force—gravity—is involved. The weight simply changes the equilibrium point about which the mass oscillates.

5. TRUE or FALSE: The motion of a simple pendulum is necessarily a simple harmonic motion.

6. The mass of the string is usually neglected in treating the motion of a simple pendulum. If the mass of the string is not completely negligible, how is the motion of the pendulum affected?

Try It Yourself #5

A certain mass hanging in equilibrium on the end of a vertical spring stretches the spring by 2.00 cm. (a) If the mass is then pulled down 5.00 cm farther and released, at what frequency does it oscillate? (b) At what speed is it moving as it passes through its equilibrium position?

Picture: The hanging mass is in equilibrium when the spring is stretched 2.00 cm. Use this to calculate k/m, which can then be used to calculate the oscillation frequency. When passing through equilibrium, the mass has its maximum speed.

Solve:

In equilibrium, the net force on the mass is zero. Apply Newton's second law to the mass in equilibrium to find an *algebraic* expression for k/m. You may wish to draw a free-body diagram of the mass in its equilibrium position to assist you. There should be two forces acting on the mass: the spring force and Earth's gravitational force.	
Use your expression from above to calculate the frequency of oscillation.	$f = 3.52$ Hz

The velocity of a particle experiencing simple harmonic motion is the time derivative of its position. Write an expression for the y position with respect to the equilibrium position as a function of time and take its time derivative *algebraically*. The amplitude is given in the problem, and you just calculated the frequency. What is the maximum value of $\sin \omega t$? Use all this to solve for the maximum speed.	
	$v_{max} = 1.11$ m/s

Check: This seems to be a reasonable speed.

Taking It Further: Where is the potential energy of this mass–spring system a maximum? Why?

Try It Yourself #6

(a) How long is a simple pendulum whose period is 2.00 s? Pendula with a period of 2 s are common in tall clocks. (b) If this pendulum has a mass of 0.200 kg and a total energy of 0.0200 J, what is the amplitude of its oscillation?

Picture: The length of a simple pendulum can be determined from the period and the local gravitational field. The amplitude of oscillation can be determined by its kinetic energy.

Solve:

Find the length of the pendulum using the expression that relates the period to its length	
	$\ell = 0.994$ m
Write an *algebraic* expression for the speed of the mass as a function of time. Determine the maximum speed of the mass.	

The maximum kinetic energy, which occurs when the speed above is a maximum, is equivalent to the total energy of the system. Use this relationship to find the amplitude of oscillation.	
	$A = 0.142$ m

Check: The amplitude of this oscillation is less than 10% of the length of the pendulum, so the result seems reasonable.

Taking It Further: Where does the maximum acceleration of this pendulum bob occur?

14.4   Damped Oscillations

In a Nutshell

In real oscillations, the motion is not conservative; it is always **damped** by frictional forces. In this chapter we consider only **linearly damped** systems, for which $\vec{F}_d = -b\vec{v}$. Here b is the damping constant. As a result, the energy and amplitude of the motion decrease with time. If damping forces are small, the energy decreases exponentially, that is, by a constant fraction in a given time interval.

The position of a damped harmonic oscillator is given by $x = A_0 e^{-(b/2m)t} \cos(\omega' t + \delta)$. The resonant frequency of a damped harmonic oscillator is less than that of an undamped oscillator: $\omega' = \omega_0 \sqrt{1 - \left(\dfrac{b}{2m\omega_0}\right)^2}$. If the damping coefficient is less than $b_c = 2m\omega_0$, then the system oscillates while its amplitude decays. It is said to be **underdamped**. If $b = b_c$, the system is **critically damped** and the object returns to equilibrium without oscillation very rapidly. If $b > b_c$, the system is **overdamped**. In this case the system returns very slowly to its equilibrium position.

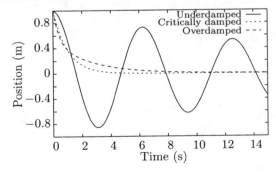

The quality factor Q of a damped oscillation expresses the amount of damping; a high Q means the oscillation takes a long time to die out. It can be interpreted as the ratio of the energy lost per cycle divided by the energy of the oscillator.

Physical Quantities and Their Units

Damping coefficient b Units of kg/s

Important Derived Results

Linear damping force $\vec{F}_d = -b\vec{v}$

Position as a function of time for a damped oscillator $x = A_0 e^{-(b/2m)t} \cos(\omega' t + \delta)$

Frequency of a damped oscillator $\omega' = \omega_0 \sqrt{1 - \left(\dfrac{b}{2m\omega_0}\right)^2} = \omega_0 \sqrt{1 - \dfrac{1}{4Q^2}}$

Q factor for a damped oscillator $Q = 2\pi \dfrac{E}{|\Delta E|}$

Try It Yourself #7

A child on a swing swings back and forth once every 3.75 s. The total mass of the child and the swing seat is 38.0 kg. At the bottom of the swing's arc, the child is moving at 1.82 m/s. (a) What is the total energy of the swing and the child if the gravitational potential energy is zero at the bottom of the swing? (b) If the Q factor is 25, what average power must be supplied to the swing to keep it moving with a constant amplitude? (Neglect the mass of the ropes.)

Picture: The child and swing can be modeled as a simple pendulum. The total energy is equal to the maximum kinetic energy of the child and swing. The Q factor is related to the fraction of energy dissipated in each cycle. The average power, energy dissipated per cycle, and period are also related. The average power supplied must equal the rate of energy dissipation in the swing.

Solve:

Find the total energy of the swing and child. The child's maximum speed is at the bottom of the arc.	
	$E_{\text{total}} = 62.9$ J
Write an *algebraic* expression for the Q factor of the oscillator in terms of the energy and energy loss per cycle.	

By multiplying the numerator and denominator of the result in the previous step by $1/T$, where T is the period, you now have an expression that includes power loss in the denominator. This power loss must be equivalent to the power the child must deliver to the swing. Rearrange and solve for P.	
	$P_{\text{av}} = 4.22$ W.

Check: This is a reasonably small power, so even a child should be able to produce this.

Taking It Further: If friction and wind resistance could be reduced, would the Q increase or decrease? What about the period? The resonant frequency? Why?

Try It Yourself #8

A certain damped harmonic oscillator loses 5% of its energy in each full cycle of oscillation. By what factor must the damping constant be changed in order to damp it critically?

Picture: The critical damping constant is related to the mass and period of the motion. The system's Q factor is also related to these parameters, so could be used to solve for b. Take the ratio to find the factor asked for.

Solve:

Find an *algebraic* expression for the critical damping constant.	
Find an *algebraic* expression for the Q factor in terms of the energy loss.	
Find an *algebraic* expression for the Q factor in terms of the damping constant, the mass, and the resonant frequency.	

Equate the two expressions for the Q factor and re-arrange to solve for the current damping constant.	
Find the ratio of the critical damping factor to the current damping factor.	$b_c/b = 251$

Check: If the system only loses 5% of its energy in each full cycle, the system must be under-damped, so the damping will have to increase to critically damp the oscillator. Our result agrees with that assessment.

Taking It Further: Once the system is critically damped, how many oscillations occur before the oscillator approaches its equilibrium configuration?

14.5   Driven Oscillations and Resonance

In a Nutshell

Here we consider the motion of an oscillator that is driven by a repetitive (periodic) driving force. Consideration is restricted to the steady-state motion that results when the energy per cycle transferred by the driving force equals that dissipated by the frictional forces. This is simple harmonic motion at the frequency of the driving force.

The Q factor of a system can also be written in term of the ratio of the resonant, or natural frequency of the oscillator to the width of the resonance curve.

Important Derived Results

Position of a driven damped oscillator	$x = A \cos(\omega t - \delta)$
Amplitude of a driven damped oscillator	$A = \dfrac{F_0}{\sqrt{m^2(\omega_0^2 - \omega^2)^2 + b^2\omega^2}}$
Phase constant of a driven damped oscillator	$\delta = \dfrac{b\omega}{m(\omega_0^2 - \omega^2)}$

Try It Yourself #9

Oscillator 1 has a peak resonance angular frequency of 19 500 Hz and a Q factor of 20. Oscillator 2 has a peak resonance angular frequency of 20 000 Hz and a Q factor of 100. On the same graph sketch an estimate of the resonance curve for each oscillator showing the power delivered from a driving source versus the angular frequency of the source. On these sketches let the peak average power for each oscillator be the same.

Picture: The linewidth of each resonance peak can be found from the Q value and the resonance frequency of each oscillator.

Solve:

Find the linewidth of oscillator 1.	
Find the linewidth of oscillator 2.	
Sketch the resonant peak power as a function of frequency	

Check: A higher Q factor should result in a narrower linewidth.

Taking It Further: The sketched answer to this problem is with the other answers in the back of this book.

QUIZ

1. TRUE or FALSE: The period of a damped oscillator is always shorter than it would be if there were no damping.

2. TRUE or FALSE: An object swinging on the end of a massless string as a simple pendulum oscillates with a period that is independent of its mass.

3. A mass oscillates on the end of a certain spring at a frequency f. The spring is cut in half, and the same mass is set oscillating on one of the pieces. What is its new frequency of oscillation?

4. A mass on the end of a string, which is hung over a nail, is set swinging as a simple pendulum of length L. While it is swinging, the string is paid out until the free-swinging length is $2L$. What happens to the pendulum's frequency? Can we use the conservation of energy to find out what happens to its amplitude? Neglect friction at the nail.

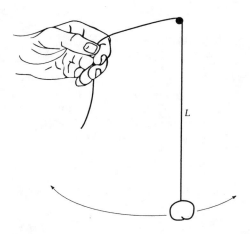

5. Would it have made any difference in our discussion in this chapter if we had defined simple harmonic motion as motion that obeys $x = A\sin(\omega t + \delta)$ rather than defining it in terms of the cosine?

6. The rim of a 0.300-kg bicycle wheel 95.0 cm in diameter is hanging from a horizontal nail on the wall. If it is knocked slightly to one side and starts swinging, what is the period at which it oscillates?

7. A phonograph record 30.0 cm in diameter revolves once every 1.80 s. (a) What is the linear speed of a point on its rim? (b) What is its angular speed? (c) Write an equation for the y component of the point's position as a function of time, given that it is at its maximum value on the y axis when $t = 0$.

Chapter 15

Traveling Waves

15.1 Simple Wave Motion

In a Nutshell

Wave motion is the transport of energy and momentum from one point to another without the transport of matter. Mechanical waves require a material medium through which to propagate (travel), whereas electromagnetic waves can propagate in a vacuum. For mechanical waves, a disturbance in a medium propagates because of the elastic properties of the medium.

A wave in which the disturbance of the medium is perpendicular to the direction of propagation is called a **transverse wave**. Waves on a plucked string are examples of transverse waves.

A wave in which the disturbance of the medium is parallel to the direction of propagation is called a **longitudinal wave**. Sound waves are longitudinal waves. Surface waves on water are a combination of the two types of waves.

If one end of a string under tension is given a sideways flip, a **wave pulse** travels down the string. Energy and momentum are carried along the string, but the material of the string itself twitches only from side to side. The speed at which the pulse travels depends on the tension in the string and its mass density μ (mass per unit length). The size, shape, and location of a pulse are described by a **wave function** $y = f(x - vt)$, where y is the displacement of the string and v is the speed of the pulse. The plus sign is used to describe a pulse moving in the negative x direction, whereas the minus sign is for one moving in the positive x direction.

For a pressure wave in a fluid, like sound traveling through air, v depends on the bulk modulus of the material B and the volume mass density ρ.

For sound waves in a gas such as air, the speed v of the wave depends on the temperature of the gas in kelvins, the universal gas constant R, the molar mass M of the gas, and a parameter γ, which depends on the molecular structure. For a monatomic gas $\gamma = 1.67$ and for a diatomic gas $\gamma = 1.40$.

All wave functions for propagating waves or pulses must satisfy the wave equation.

Physical Quantities and Their Units

Molar mass for air $\qquad\qquad\qquad M = 29.0 \times 10^{-3}$ kg/mol

Universal gas constant $\qquad\qquad R = 8.3145$ J/mol $\cdot$ K

Gas constants $\qquad\qquad\qquad\quad \gamma = 1.67$ (monatomic gas)

$\qquad\qquad\qquad\qquad\qquad\quad \gamma = 1.40$ (diatomic gas)

Fundamental Equations

Wave equation
$$\frac{\partial^2 y}{\partial x^2} = \frac{1}{v^2}\frac{\partial^2 y}{\partial t^2}$$

Important Derived Results

Speed of waves or pulses on a string
$$v = \sqrt{\frac{F_T}{\mu}}$$

Speed of sound waves in a liquid
$$v = \sqrt{\frac{B}{\rho}}$$

Speed of sound waves in a gas
$$v = \sqrt{\frac{\gamma RT}{M}}$$

Common Pitfalls

> The term "amplitude" has the same meaning for a sinusoidal mechanical wave as it does for simple harmonic motion. It is the maximum displacement of a portion of the medium from its undisturbed, equilibrium position. It is *not* the distance from crest to valley.

> Transverse waves propagating along a string disturb the string; that is, they displace it. Be careful not to confuse the transverse velocity of a segment of the string due to the wave with the velocity of the wave disturbance itself. The velocity of the wave is the rate at which the pattern of disturbance is moving along the string. The two velocities are not directly related and, in fact, are directed at right angles with each other. (In a longitudinal wave the two velocities are in opposite directions 50% of the time.)

1. TRUE or FALSE: The speed at which sound waves propagate in a material medium depends upon the speed of the sound source relative to the medium.

2. A Slinky is a children's toy that is just a long, loose-coiled spring. It is quite useful as a wave demonstrator. How would you generate longitudinal waves in a Slinky? What about transverse waves?

Try It Yourself #1

You stand on the edge of a canyon 420 m wide and clap your hands. An echo comes back to you 2.56 s later. What is the temperature of the air? For air, $\gamma = 1.40$.

Picture: The speed of the wave can be found from the distance and time provided. Remember that the sound has to go across the canyon and back. Use the formula for the speed of sound waves in a gas in terms of temperature, molar mass, and γ to find the temperature.

Solve:

Calculate the speed of the wave from the kinematic quantities given.	
Find the temperature using the expression for the speed of sound in a gas.	
	$T = 268$ K $= -4.9°$C

Check: Although maybe a bit brisk, this is certainly a reasonable temperature in which to be standing outside.

Taking It Further: Assuming you knew the temperature, but nothing about the composition of the air, could you use this experiment to learn anything about the constituents of the air?

15.2 Periodic Waves

In a Nutshell

If one end of a taut string is driven transversely in periodic motion (displacement is a sinusoidal function of time), then a **periodic wave** is generated. If a periodic wave is traveling along a taut string or any other medium, each point along the medium oscillates with the same period.

If a **harmonic wave** is traveling through a medium, each point of the medium oscillates in simple harmonic motion. When a harmonic wave travels along the string, the shape of the string at any particular instant in time is that of a sine function. The distance between two successive wave crests is called the **wavelength** λ. The wavelength is the distance after which the shape repeats itself. When a harmonic wave propagates along a string, each point on the string moves in simple harmonic motion with the same frequency f and **period** $T = 1/f$. During one period, the wave crest moves a distance equal to the wavelength. This gives the relationship for the **speed of a wave**: $v = \lambda f$.

The wave function describing a harmonic wave traveling in the $+x$ direction with speed v is $y(x,t) = A \sin(kx - \omega t)$, where A is the **amplitude** of the wave, $\omega = 2\pi f$ is the **angular frequency** of the wave, and $k = 2\pi/\lambda$ is the **wave number** of the wave. The argument of the sine function is the **phase**.

Even though matter is not transported by a wave, energy is. The rate of energy transfer (power) at a point on a taut string (which supports transverse motion of the string) is given by $P = F_{Ty}v_y \approx -F_T(\partial y/\partial t)(\partial y/\partial x)$, where F_T is the tension in the string, and the wave propagates along the $+x$ direction. For a harmonic wave the average power and average energy are both proportional to the amplitude squared.

In **harmonic sound waves** the molecules of the medium undergo simple harmonic motion longitudinally, which means the pressure varies sinusoidally in space and time. The pressure is $90°$ out of phase with the displacement of the molecules.

Physical Quantities and Their Units

Period	dimensions T, units of s
Frequency	dimensions 1/T, units of 1/s
Wave number	dimensions 1/L, units of rad/m

Fundamental Equations

Period	$T = \dfrac{1}{f}$
Angular frequency	$\omega = 2\pi f = \dfrac{2\pi}{T}$
Wave number	$k = \dfrac{2\pi}{\lambda}$
Wave speed	$v = \lambda f = \dfrac{\lambda}{T} = \dfrac{\omega}{k}$

Important Derived Results

Harmonic wave function	$y(x,t) = A\sin(kx - \omega t + \delta)$
Transverse wave velocity	$v_y = \dfrac{\partial y}{\partial t} = -\omega A\cos(kx - \omega t + \delta)$
Power of wave on a string	$P = \mu v \omega^2 A^2 \cos^2(kx - \omega t + \delta)$
Average power of wave on a string	$P_{av} = \frac{1}{2}\mu v \omega^2 A^2$
Energy in wave on a string	$(\Delta E)_{av} = \frac{1}{2}\mu \omega^2 A^2 \, \Delta x$
Harmonic sound wave	$s(x,t) = s_0 \sin(kx - \omega t + \delta)$
Pressure of sound wave	$p = -p_0 \cos(kx - \omega t + \delta)$
Pressure amplitude	$p_0 = \rho \omega v s_0$
Energy of sound wave	$(\Delta E)_{av} = \frac{1}{2}\rho \omega^2 s_0^2 \, \Delta V$

Common Pitfalls

➤ The equation $v = f\lambda$ is really just another way of stating that the speed equals the distance divided by the time. The wavelength is the distance and the frequency is the reciprocal of the period—the time it takes the disturbance to travel one wavelength. The wave speed v is determined by the properties of the medium that transports the wave and may or may not depend on the wave's frequency. The speed of sound in air, for example, is nearly independent of the frequency, whereas the speed of ripples on a pond depends strongly on the frequency. When the speed of a wave is independent of frequency, the wavelength changes accordingly to keep the speed constant.

➤ "Phase" refers to the entire argument of the sine function at any time t and position x. "Phase constant" refers specifically to the phase at time $t = 0$ and position $x = 0$.

3. TRUE or FALSE: Every element of a string on which a harmonic wave is propagating is undergoing simple harmonic motion.

4. Why, in a medium transporting a harmonic sound wave, is the pressure at a given point 90° out of phase with the displacement at that point?

Try It Yourself #2

The temperature inside a house is maintained at 20.0°C, so the speed of sound in the air within the house is 340 m/s. On a cold day the speed of sound in the air outside the house is 334 m/s. A tuning fork in the house is struck and starts vibrating at 440 Hz. A boy outside the house hears the sound from the tuning fork. What frequency does he hear? Determine the wavelength of the wave both in the air inside the house and in the air outside the house.

Picture: Determine the frequency with which the sound wave enters the cold air. Determine the frequency of the sound in the cold air. Wavelength can be found from the wave speed and frequency.

Solve:

What frequency does the wave have when it hits the warm air/cold air interface?	
This now serves as the source of the wave in the cold air. What is the frequency of the wave in the cold air?	
	$f = 440$ Hz
Find the wavelength of the wave in the house.	
	$\lambda_{\text{warm}} = 0.773$ m

Find the wavelength of the wave outside.	
	$\lambda_{\text{cold}} = 0.759$ m

Check: Since the frequency is the same in both cases, the wavelength should be smaller when the speed is smaller.

Taking It Further: Does the pitch of the sound the boy hears increase or decrease as he walks inside from the cold? Explain.

Try It Yourself #3

The wave function for a harmonic wave is $y(x,t) = 0.002\sin(31.4x - 628t + 1.57)$, where y and x are in meters and t is in seconds. (a) Find the amplitude, wavelength, frequency, period, and phase constant of this wave. (b) In what direction does this wave travel and what is its speed? (c) At $t = 0.0200$ s find the phase at $x = 0.300$ m.

Picture: Use the expression of the harmonic wave to identify the terms in the wave function. The direction of travel can be determined from the relative sign of the position and time terms in the argument of the sine function. The phase is the argument of the sine function.

Solve:

Compare the general formula of a harmonic wave with the function given to find the appropriate parameters.	
	$A = 0.002$ m, $\lambda = 0.200$ m, $f = 100$ Hz, $T = 0.01$ s, $\delta = 1.57$ rad
The direction of the wave is in the positive direction because the position and time terms have opposite signs. The speed can be found from the wavelength and frequency.	
	$v = 20.0$ m/s

The phase is the argument of the sine function. Find the phase for the given time.	
	$\theta = -1.57$ rad

Check: Make sure you have taken care of all the factors of 2π, and watch your units. It may help to remember that the argument of the sine function has units of rad, so all those factors of 2π also have units of rad.

Taking It Further: Is this a longitudinal or transverse wave? How can you tell?

15.3 Waves in Three Dimensions

In a Nutshell

When a pebble is dropped onto the surface of a pond, a few surface waves ripple outwardly in a circular pattern. When a small, steadily vibrating object is placed on the surface of the pond, a steady train of surface waves will ripple outwardly in a circular pattern. When the vibrator is immersed deeply in the water, the steady train of compression maxima and minima produced will ripple outward in a spherical pattern. A spherical surface moving outward with one of these compression maxima is called a **wavefront**. A wavefront is a surface of constant phase. The outward motion of the wavefronts can be illustrated by **rays**, which are lines directed perpendicularly to the wavefronts. For the spherical wavefronts described here, the rays are straight lines directed away from the wave source.

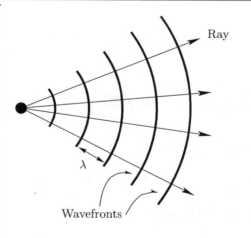

For a **plane wave** the rays are parallel to each other. At large distances from a point source, a small region of a spherical wave can be approximated by a plane wave.

If a wave source transfers energy at an average rate of P_{av}, some distance away we measure the **intensity** I of the wave. The intensity is the average power of the source per unit area that is perpendicular to the direction of propagation. I is often measured in units of watts per square meter. If the source is surrounded by a sphere of radius r, centered at the source, then the area is the area of a sphere, $A = 4\pi r^2$. So we can see that the intensity decreases as $1/r^2$ the farther we get from the source.

The sensation of loudness depends upon both intensity and frequency. However, the dependence of loudness on frequency will not be considered here. The sensation of loudness is found to vary more or less logarithmically with intensity, so the **intensity level** β is expressed in decibels (dB): $\beta = (10 \text{ dB}) \log_{10} I/I_0$, where I is the intensity of a sound and I_0 is a reference level that we take to be the threshold of hearing, $I_0 = 10^{-12} \text{ W/m}^2$.

The threshold of pain occurs at an intensity of about 1 W/m^2. As the intensity varies from 10^{-12} W/m^2 to 1 W/m^2, the intensity level varies from 0 to 120 dB. A doubling of intensity results in a 3-dB increase in intensity level.

Physical Quantities and Their Units

Intensity I	units of W/m^2
Average energy density η	units of J/m^3
Intensity level β	units of dB
Threshold of hearing	$I_0 = 10^{-12} \text{ W/m}^2$; $\beta = 0$ dB
Pain threshold for hearing	$I = 1 \text{ W/m}^2$; $\beta = 120$ dB

Fundamental Equations

Intensity	$I = \dfrac{P_{\mathrm{av}}}{A}$
Intensity level in dB	$\beta = (10 \text{ dB}) \log_{10} \dfrac{I}{I_0}$

Important Derived Results

Intensity of a point source	$I = \dfrac{P_{\mathrm{av}}}{4\pi r^2}$
Intensity of harmonic sound wave	$I = \eta_{\mathrm{av}} v = \frac{1}{2}\rho\omega^2 s_0^2 v = \dfrac{1}{2}\dfrac{p_0^2}{\rho v}$

Common Pitfalls

> ➤ Although the terms "intensity" and "intensity level" are similar, they represent quite different quantities. The intensity I is the power per unit area, whereas the intensity level β equals $(10 \text{ dB}) \log_{10} I/I_0$, where I_0 is a reference intensity. It can be useful to memorize that 3 dB refers to an intensity ratio of 2.

> ➤ To simplify physics problems we usually assume that a sound source is isotropic—that is, the intensity is broadcast uniformly in all directions. As you know, actual sound sources, such as a loudspeaker or a human voice, generally have intensities that are greater in the forward direction.

5. TRUE or FALSE: The decibel is a unit of intensity.

6. Consider a sound wave that propagates uniformly in all directions from a point source. How does the amplitude of the wave vary with distance from the source?

Try It Yourself #4

Suppose each of the two loudspeakers of my stereo system is delivering 1.00 W of average power in the form of audible sound in my office. Also suppose that this power is being delivered uniformly in all directions. (a) If I am sitting 2.5 m away from the speakers, what is the intensity of the sound I hear directly from the speakers (that is, neglecting the sound reflected from the walls of the room)? (b) To what intensity level does this correspond?

Picture: Calculate the direct sound intensity from the speakers using their power output and the receiver's distance from them. Convert the calculated intensity to an intensity level in dB.

Solve:

Calculate the intensity from one of the speakers. Because the two loudspeakers are not coherent (see Chapter 16), the intensity for two speakers is twice the intensity for one speaker.	$I_{\text{tot}} = 2.54 \times 10^{-2}$ W/m^2
Find the intensity level.	$\beta = 104$ dB

Check: This is an extremely loud sound that would likely chase you out of the room. You don't buy amplifiers that put out lots of power because you need that much acoustical power; rather, it's because good-quality loudspeakers do not convert electrical power to acoustical power very efficiently.

Taking It Further: How would you determine what output power of a single speaker would produce a nice ambient background level of music?

Try It Yourself #5

Assume that a barking dog can put out (on the average) 1 mW of sound power and that the sound propagates uniformly in all directions. What is the sound intensity level (in dB) 6 m away from a pen containing eight barking dogs?

Picture: Calculate the direct sound intensity from the dogs using their power output and the receiver's distance from them. Convert the calculated intensity to an intensity level in dB.

Solve:

Calculate the intensity from the dogs.	
Find the intensity level.	$\beta = 72$ dB

Check: This intensity level is comparable to standing on the sidewalk of a busy street.

Taking It Further: Why, then, are a neighbor's barking dogs often much more annoying than the traffic right outside your door?

15.4   Waves Encountering Barriers

In a Nutshell

When a wave is incident on a boundary that separates two regions of different wave speed, part of the wave is reflected and part is transmitted. A pulse on a string that is attached to a more dense string is reflected at the boundary as an inverted wave (an inversion corresponds to a phase shift of 180°). However, if the second string is less dense than the first, the reflected pulse is not inverted. In either case, the transmitted pulse is not inverted. If the string is tied to a fixed point, the reflected pulse is inverted.

When a sound wave traveling in air strikes a solid or liquid surface, the reflected and incident rays make equal angles with the normal to the boundary, and the angle between the normal and the transmitted ray differs from that between the normal and the incident ray. This change of direction (bending) of the transmitted ray is called **refraction**. The angle that the refracted ray makes with the normal is greater or less than that of the incident ray, depending on whether the wave speed in the second region is greater or less than the wave speed in the incident medium.

When a portion of a wavefront is blocked by an obstruction, the portions that are not blocked will spread out in all directions—including into the region behind the obstruction. This spreading out, which in principle can be observed whenever a part of a wavefront is blocked, is called **diffraction**. Almost all the diffraction occurs for that part of the wavefront that passes within a few wavelengths of the edge of the obstacle. For the parts of the wavefront that pass more than a few wavelengths from the edge, diffraction is negligible and the wave propagates in straight lines in the direction of the incident rays.

When a wave travels through an aperture, the amount of diffraction depends on the size of the aperture relative to the size of the wavelength. If the wavelength is large relative to the aperture size, the waves spread out as they pass through the aperture, as if the waves were originating from a point source. If the wavelength is small, the waves pass straight through with almost no diffraction.

Important Derived Results

Reflection coefficient for wave on a string	$r = \dfrac{h_r}{h_{in}} = \dfrac{v_2 - v_1}{v_2 + v_1}$
Transmission coefficient for wave on a string	$\tau = \dfrac{h_t}{h_{in}} = \dfrac{2v_2}{v_2 + v_1}$
Energy conservation	$1 = r^2 + \dfrac{v_1}{v_2}\tau^2$

Common Pitfalls

7. TRUE or FALSE: A wave pulse reflected at the end of a string is always inverted.

8. Why is it that we can hear, but not see, around corners?

Try It Yourself #6

A 2.00-m-long piece of heavy string, with a mass of 75.0 g, is tied to 3.00 m of light twine with a mass of 25.0 g, and the combination is put under a tension of 50.0 N. If a transverse pulse is sent down the heavy string, determine the reflection and transmission coefficients at the junction point.

Picture: The reflection and transmission coefficients depend on the speed of the wave in the two string segments.

Solve:

Determine the speed of waves on the heavy string using its mass, length, and tension.	
Determine the speed of waves on the light string using its mass, length, and tension.	

Determine the reflection coefficient at the point where the two strings join.	
	$r = 0.359$
Determine the transmission coefficient at the point where the two strings join.	
	$\tau = 1.36$

Check: The speed of a wave is slower on the heavy string, so we should expect that the wave is not inverted on reflection, which our answers reveal.

Taking It Further: What can you say about the relative transmitted and reflected pulse sizes and orientations if the pulse starts on the light string instead?

15.5   The Doppler Effect

In a Nutshell

When a wave source and receiver are moving relative to each other, the rate at which the compressions leave the source is not the same as the rate at which they arrive at the receiver. Thus, the frequency observed by the receiver differs from the frequency emitted by the source. This phenomenon is called the **Doppler effect**. The observed frequency depends on the frequency of the emitted radiation f_s, the speed of the source u_s relative to the medium, the speed of the receiver u_r relative to the medium, and the speed of the wave in the medium v. If the receiver and the source are moving closer to each other, the observed frequency is higher than the source frequency. If the receiver and source are moving farther apart, then the observed frequency is lower than the source frequency.

When the relative speed of approach u of the source and receiver is small compared to the speed of sound v, the **Doppler shift** in frequency can be written as $\Delta f / f_s \approx \pm u/v$, where $u = u_s \pm u_r$ is the speed of the source relative to the receiver. We use the plus sign if the distance between the source and receiver is decreasing.

When dealing with electromagnetic radiation, there is no propagation medium, and we must use the **relativistic Doppler-effect** formula.

When a source moves at a speed greater than the speed of the wave, a conical **shock wave** with the source at the apex forms behind the source at an angle θ with the path of the source. The angle θ between one side of the cone and the path of the source is related to the speed u of the source by the expression $\sin \theta = v/u$.

The ratio of the source speed u to the wave speed v is called the **Mach number**.

Solving Problems Involving Doppler Shift

Picture: Solving problems involving the Doppler shift means using the Doppler-shift equation:

$$f_{\mathrm{r}} = \frac{v \pm u_{\mathrm{r}}}{v \pm u_{\mathrm{s}}} f_{\mathrm{s}}$$

Solve:

1. Find the speed of the source u_{s} and the receiver u_{r} in the reference frame of the propagation medium.
2. Find the directions of the motions of the source and receiver in the same reference frame.
3. Substitute values into the Doppler-shift equation. Both the source moving toward the receiver and the receiver moving toward the source tend to increase the received frequency. Thus, if the source is moving toward the receiver choose the minus sign in the denominator, and if the receiver is moving toward the source, choose the plus sign in the numerator.
4. If the wave bounces off a reflector before reaching the receiver, treat the reflector first as a receiver and apply the Doppler-shift equation, then treat the reflector as a source and apply the Doppler-shift equation once again.

Check: If the distance between a source and receiver is decreasing, then the received frequency f_{r} is higher than the source frequency f_{s}. If the distance is increasing, then f_{r} is lower than f_{s}.

Important Derived Results

Doppler effect	$f_{\mathrm{r}} = \dfrac{v \pm u_{\mathrm{r}}}{v \pm u_{\mathrm{s}}} f_{\mathrm{s}}$
Doppler shift	$\dfrac{\Delta f}{f_{\mathrm{s}}} \approx \pm \dfrac{u}{v} \quad (u \ll v)$
Relativistic Doppler effect	$f_{\mathrm{r}} = \sqrt{\dfrac{c \pm u}{c \mp u}}\, f_{\mathrm{s}}$
Mach angle	$\sin \theta = \dfrac{v}{u}$
Mach number	$\text{Mach number} = \dfrac{u}{v}$

Common Pitfalls

> ➤ It is really easy to make sign errors when doing Doppler-effect problems. Simply memorizing formulas will not suffice. Remember, in the frequency formula given, u_{r}, u_{s} and v are *never* negative because they are speeds, not velocities. A source moving in the direction of a receiver tends to produce an increase in the observed frequency, so the minus sign preceding u_{s} is used. A receiver moving in the direction of the source tends to produce an increase in frequency, so the plus sign preceding u_{r} is used. If both the source and receiver are moving, u_{s} and u_{r} are the speeds of the source and receiver relative to the medium, not to each other.

9. TRUE or FALSE: The received frequency of a sound wave is always equal to the frequency of the source, even when there is relative motion between the source and the receiver.

10. A tuning fork, which vibrates at a fixed frequency, is being waved back and forth along the direction from it to you. Describe what you hear. What will you hear if the fork is moved from side to side?

Try It Yourself #7

On a still day, a car traveling at 100 km/h sounds its horn, which has a frequency of 250 Hz. What frequency is heard by a receiver who is proceeding at 60.0 km/h in the same direction (a) directly ahead of the car and (b) directly behind it?

Picture: Both the source and the receiver are moving. In part (a), the source is moving toward the receiver, which will cause an upward shift in the received frequency. Choose the correct sign in the denominator for this to happen. The receiver, however, is moving away from the source, which will cause a downward shift in the received frequency. Choose the correct sign in the numerator to provide this downward shift. In part (b), the source is moving away from the receiver and the receiver is moving toward the source. Use the signs that tend to provide the appropriate frequency shifts. The speed of sound is 340 m/s.

Solve:

Convert all speeds to m/s.	
The speeds of the wave, source, and receiver are all given. Apply the Doppler-effect equation to find the received frequency for the receiver ahead of the car.	$f_{\text{front}} = 259$ Hz
The speeds of the wave, source, and receiver are all given. Apply the Doppler-effect equation to find the received frequency for the receiver behind the car.	$f_{\text{back}} = 242$ Hz

Check: For part (a), the net effect of the motion is that the source is moving toward the receiver, so we expect the received frequency to be higher than the source frequency. Exactly the opposite is true for part (b).

Taking It Further: What will happen to the observed frequency of the horn when the 100 km/h car passes the car ahead of it?

Try It Yourself #8

On a still day, a car moves at a speed of 40.0 km/h directly toward a stationary wall. Its horn emits a sound at a frequency of 300 Hz. (a) If the velocity of sound is 340 m/s, find the frequency at which the sound waves hit the wall. (b) The waves reflect off the wall and are received by the driver of the car. What frequency does she hear?

Picture: Work this problem in steps. Find the frequency of the sound "received" by the wall. Find the frequency of the sound the driver hears after the wave has bounced off the wall. Remember that in this part the "receiver" is moving.

Solve:

Convert the speed of the car to m/s.	
Find the received frequency at the wall. Because the car is moving toward the wall, the frequency at the wall will be greater than the source frequency. The wall, however, is not moving.	$f_{r, \text{ wall}} = 310$ Hz
Find the frequency the driver hears. The frequency that hits the wall is reflected at that same frequency, so the wall acts like a source that is not moving, emitting the frequency found above. Because the car is moving toward the source, the frequency heard will again be greater than the source frequency.	$f_{r, \text{ car}} = 320$ Hz

Check: Both received frequencies are, in fact, larger than the source frequencies, as expected.

Taking It Further: Can you envision a situation in which the driver will hear a lower frequency than that emitted by her horn? How could this be accomplished?

QUIZ

1. TRUE or FALSE: A wave passing through a hole whose diameter is less than the wavelength propagates outward from the hole in all directions on the far side of the barrier, as though from a point source.

2. TRUE or FALSE: If two elements of a string on which a harmonic wave is propagating move in phase, the two elements must be an integral number of wavelengths apart.

3. A nice, pleasant sound level at which to listen to music is 60 dB. Would 120 dB be twice as loud?

4. Imagine that you are listening to orchestral music being played on a stereo that is in another room down the hall. Will diffraction have any effect on the sound you hear?

5. For a function to be a wave function it must satisfy certain criteria. One necessary criterion is that it satisfy the condition $y(x,t) = y_1(x - vt) + y_2(x + vt)$. A string must be continuous. Therefore only continuous functions can be wave functions for waves on a string. Which of the following functions can be wave functions for waves on a string?

    (a) $y(x,t) = Ae^{-(x-vt)^2/2a^2}$

    (b) $y(x,t) = Ae^{-x^2}e^{+bt}$

    (c) $y(x,t) = \begin{cases} 0 & \text{for } x \leq vt - a \\ D & \text{for } vt - a \leq x \leq vt + a \\ 0 & \text{for } x \geq vt + a \end{cases}$

6. My car's horn sounds a tone of frequency 250 Hz. If I am driving directly at you at 25.0 m/s, blowing my horn, on a hot, still day (93.0°F), (a) what is the wavelength of the sound that reaches you? (b) What frequency do you hear?

7. A woman with perfect pitch, standing next to a railroad track on a still day, is amused to notice that the whistle of an approaching train is sounding a true concert A (440 Hz). After the train has passed her and is receding, the pitch has dropped to a true F natural (349 Hz). (a) How fast was the train going? (b) What was the actual frequency of the whistle?

Chapter 16

Superposition and Standing Waves

16.1 Superposition of Waves

In a Nutshell

The **principle of superposition** states that when two or more waves overlap, the resultant wave is the algebraic sum of the individual waves. That is, $y(x,t) = y_1(x,t) + y_2(x,t)$. This principle holds for any medium as long as the proportional limit of the medium is not exceeded.

The superposition of harmonic waves is called **interference**. Consider two harmonic waves of equal amplitude and frequency that are simultaneously traveling along the same string. If the crests of one of these waves exactly superpose the crests of the other, the difference between the phases of the two waves is zero. Such waves are said to be in phase and the resultant wave will have an amplitude equal to the sum of the amplitudes of the original waves. This phenomenon is called **constructive interference**. On the other hand, when the crests of one of these waves exactly superpose the troughs of the other, the difference between the phases of the two waves is 180° (π rad) and the resultant wave has an amplitude equal to zero—the difference between the amplitudes of the original waves. This is called **destructive interference**.

For interference to occur, the waves do not have to be either in phase or 180° out of phase. We will, however, only consider waves with a phase difference that remains fixed (is constant).

When two harmonic waves with slightly different frequencies interfere, **beats** occur. We experience this in sound waves when two slightly different frequencies f_1 and f_2 interfere. Addition of the harmonic waves results in a sound with the *average* frequency $(f_1 + f_2)/2$, which fades in and out with the **beat frequency** $f_{\text{beat}} = |\Delta f| = |f_1 - f_2|$.

A common cause of phase difference between two waves is the difference in path length between each source and the point of interest. For two waves whose sources are in phase, if the path length difference is an integer number of wavelengths, then the two waves interfere constructively.

Coherent sources are sources that vibrate so that the phase difference between the two sources remains fixed. Coherent sound sources can be produced by two loudspeakers that are driven by the same electrical signal.

Sources that vibrate so that the phase difference is not constant but varies randomly over time are called **incoherent sources**. Two violins playing the same music in an orchestra are incoherent sources.

Fundamental Equations

Superposition principle
$$y(x,t) = y_1(x,t) + y_2(x,t) + y_3(x,t) + \ldots$$

Important Derived Results

Interference of harmonic waves

$$y = y_1 + y_2$$
$$= A\sin(kx - \omega t) + A\sin(kx - \omega t + \delta)$$
$$= 2A\cos\left(\frac{\delta}{2}\right)\sin\left(kx - \omega t + \frac{\delta}{2}\right)$$

Superposition of harmonic waves at a given point

$$p = p_1 + p_2 = p_0\sin(\omega_1 t) + p_0\sin(\omega_2 t)$$
$$= 2p_0\cos\left(\tfrac{1}{2}\Delta\omega t\right)\sin\left(\omega_{\mathrm{av}}t\right)$$

Beat frequency

$$f_{\mathrm{beat}} = |\Delta f| = |f_1 - f_2|$$

Phase difference—path-length difference relation

$$\delta = (2\pi\ \mathrm{rad})\frac{\Delta x}{\lambda} = (360°)\frac{\Delta x}{\lambda}$$

Common Pitfalls

> Destructive interference means that the two interfering waves are 180° out of phase with each other and so will tend to cancel. The cancellation is total, however, only if the two waves have equal amplitudes.
> Two wave sources need not be vibrating in phase in order for interference to occur, but they must be coherent—that is, the phase difference between their motions must be constant.

1. TRUE or FALSE: For the destructive interference of two harmonic waves from two point sources that vibrate in phase to occur, the path-length difference must be an integral number of wavelengths.

2. In the figure two wave pulses (one upright and one inverted) of the same size and shape approach one another along a string under tension. A short time later, the pulses superpose (overlap). When the pulses completely superpose, destructive interference occurs and the string is momentarily flat. At the moment when the string is flat, where is the energy that the pulses were carrying? Sketch both the displacements y and velocities v_y of the string elements as a function of position x along the string at the moment when the string is flat.

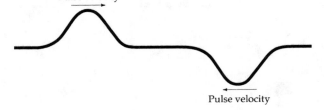

Pulse velocity

Pulse velocity

Try It Yourself #1

A string has a tension of 10.0 N and a mass per unit length of 5.00 g/m. Two harmonic waves traveling in the same direction on this string, each with a 20.0-cm wavelength and a 1.00-cm amplitude, differ in phase by 90°. Determine the amplitude and the power of the wave that results when these waves interfere.

Picture: Because both waves are on the same string and have the same wavelength, they will have the same frequency. The amplitude of the wave resulting from interference of the two harmonic waves can be expressed in terms of the phase difference between the two waves. The power depends on the angular frequency, the amplitude, the wave speed, and the linear mass density.

Solve:

Determine an *algebraic* expression for the wave function of the resultant wave by adding the two harmonic waves.	
The amplitude of the resultant wave is the time-invariant piece of the wave function.	$A_r = 1.41$ cm
Find an *algebraic* expression for the speed of the wave on the string.	
Determine an *algebraic* expression for the angular frequency of the resultant wave in terms of the wave speed and wavelength. The resultant wavelength will be the same as the wavelength of the individual waves.	
You learned how to calculate the power of a harmonic wave on a string in Chapter 15. Use that result to calculate the power.	$P = 44.1$ J

Check: Because the waves are not perfectly in phase, the resultant amplitude should be less than twice the amplitude of a single wave, which it is.

Taking It Further: How does this power compare to the power of a single wave with the given 1.00-cm amplitude? To The power of two waves with the same amplitude that didn't interfere? To the power of a single wave with twice the amplitude?

Try It Yourself #2

Two loudspeakers radiate sound waves in phase at a frequency of 100 Hz. A listener is 8.50 m from one speaker and 13.6 m from the other. Either speaker alone would produce a sound intensity level of 75 dB at the position of the listener. (a) What is the sound intensity level at the listener due to both speakers? (b) What is the intensity level at the listener if the same two speakers are 180° out of phase?

Picture: Determining the intensity level requires knowledge of the intensity, which requires knowing the net amplitude of the sound where the listener is. You will first have to determine how the sound from the two speakers is interfering, which will allow you to determine the sound amplitude, and hence the intensity level.

Solve:

Find the path-length difference of the two sound waves from the speakers to the listener.	
Find the wavelength of the sound assuming the speed of sound in air is 340 m/s.	
Compare the path-length difference to integer and half-integer values of the wavelength. What kind of interference is occurring?	
Determine the intensity level.	Because the waves interfere destructively, the intensity is zero, which cannot be expressed on the decibel scale.
For part (b), determine whether the waves interfere constructively or destructively if one speaker is 180° out of phase from the other.	

Find the amplitude of the resulting sound as compared to the amplitude of a single speaker.	
From this amplitude, calculate the new intensity as compared to the intensity from a single speaker.	
From this change in intensity, calculate a new intensity level.	81.0 dB

Check: A doubling of intensity corresponds to an increase of 3 dB. Since the intensity in part (b) is four times the intensity from one speaker, this corresponds to an increase of 6 dB in intensity level.

Taking It Further: Qualitatively, how would the intensity level change for both parts (a) and (b) if the frequency were changed to 150 Hz?

16.2   Standing Waves

In a Nutshell

Waves traveling along a taut string with one or both ends fixed reflect at the ends of the string, just as waves traveling along a column of air with one or both ends open reflect at the end of the pipe. For harmonic waves of certain specific frequencies, the waves reflecting back and forth can superpose to form a stationary vibration pattern called a **standing wave**.

When one end of the string is vibrated at a standing wave frequency, it responds "enthusiastically." This enthusiastic response is called **resonance**, and the frequencies at which the string resonates (vibrates in a standing wave) are called the **resonant frequencies** of the string. On a string that is vibrating in a standing wave there are locations called **nodes** where the string remains stationary. Between each pair of adjacent nodes is a location of maximum amplitude called an **antinode**.

The standing-wave patterns of a string that is fixed at both ends are shown here. The lowest resonance frequency for such a string is called its **fundamental frequency** f_1. When the string vibrates at its fundamental frequency, the wave pattern produced is referred to as the **fundamental mode** of vibration or the **first harmonic**. The wave pattern produced when the string vibrates at the second lowest resonance frequency f_2 is called the **second harmonic**, and the pattern when it vibrates at the nth lowest frequency f_n is the nth harmonic. For these standing waves, the resonance frequency f_n of the nth harmonic is related to the fundamental frequency by $f_n = n f_1$.

From the figure, you can see that for a flexible string fixed at both ends, the wavelength λ_n of the nth harmonic is related to the length L of the string by the equation $L = n\lambda_n/2$.

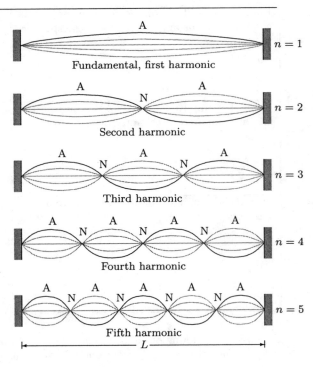

Standing sound waves have much in common with the standing waves of a vibrating string. If air is confined in a tube of length L with both ends open, any standing waves in the air in the tube will have pressure nodes (points of atmospheric pressure), and displacement antinodes, just outside both ends of the tube. As a consequence, the standing wave condition and resonant frequencies are the same for an open tube as they are for a string fixed at both ends.

In the terminology often used in music, the frequencies of the second and higher harmonics are called **overtones**. Thus the frequency of the second harmonic is called the first overtone, that of the third harmonic is called the second overtone, and so forth. The resonant frequencies of the string are called its **natural frequencies**, and each vibrational pattern is called a **mode of vibration**. A sequence of natural frequencies that are integral multiples of a fundamental frequency is called a **harmonic series**, and the individual natural frequencies are called **harmonics**.

Standing waves on a taut string can also be produced on a string with one end fixed and the other end free. The standing-wave patterns for such a string are shown here. Of particular note is that with one end free, only the odd-numbered harmonics appear. The length of the string and the harmonic wavelengths are related by $L = n\lambda_n/4$, where n can only be an odd integer.

To realize the condition just described, we must provide a connection that maintains tension F_T in the string, yet allows the free end to oscillate up and down with negligible friction. This can occur only in an ideal world. The analogous arrangement (an open-ended organ pipe) for sound waves is more readily achieved.

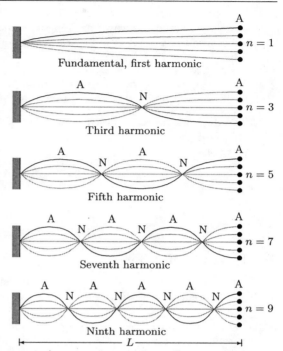

Fundamental, first harmonic $n = 1$

Third harmonic $n = 3$

Fifth harmonic $n = 5$

Seventh harmonic $n = 7$

Ninth harmonic $n = 9$

L

In a tube with only one end open to the atmosphere, the standing waves of the air in the tube have a displacement antinode just outside the open end. This displacement antinode is also a pressure node since the pressure at the open end is fixed at the pressure throughout the room, usually atmospheric pressure. Thus, for a standing wave in such a tube, the standing wave condition and resonant frequencies are the same as for a string fixed at one end.

Important Derived Results

Standing-wave condition, both ends fixed (closed) or both ends free (open)
$$\lambda_n = \frac{2L}{n} \qquad n = 1, 2, 3, \ldots$$

Resonance frequencies, both ends fixed (closed) or both ends free (open)
$$f_n = \frac{v}{\lambda_n} = n\frac{v}{2L} = nf_1 \qquad n = 1, 2, 3, \ldots$$

Standing-wave condition, one end fixed (closed) and one end free (open)
$$\lambda_n = \frac{4L}{n} \qquad n = 1, 3, 5, \ldots$$

Resonant frequencies, one end fixed (closed) and one end free (open)
$$f_n = \frac{v}{\lambda_n} = n\frac{v}{4L} = nf_1 \qquad n = 1, 3, 5, \ldots$$

Common Pitfalls

> The fundamental frequency of a taut string is the lowest frequency at which it can resonate. Correspondingly, the fundamental has the longest wavelength of any of the resonance frequencies—that is, the higher harmonics have shorter wavelengths.

> The frequencies of a vibrating string are calculated as if there were nodes at both ends. Clearly this doesn't accurately describe the usual physics laboratory experiment in which one of the ends is being driven in simple harmonic motion. However, the driven end is very nearly a node if the amplitude of the standing wave is large compared to that of the driven end.

> A vibrating string, such as a guitar string or a piano string, can act as a source of sound waves of the same frequency in air as on the string. But notice that the wavelength of the sound waves and the wavelength of the standing waves on the string are not the same. This is because the wave speeds for transverse waves on the string and for sound waves in air are different.

3. TRUE or FALSE: When a string is vibrating in a standing wave pattern, the motion at one antinode is in phase with the motion at all other antinodes.

4. The design of most orchestral instruments is based on the resonant frequencies of either a stretched string or air in a pipe. What happens to these frequencies as the temperature increases?

Try It Yourself #3

Three successive resonant frequencies of a taut string are 273, 364, and 455 Hz. Determine the fundamental frequency of the string.

Picture: We do not know if the string is fixed at both ends, or simply one end. If it is fixed at both ends, the fundamental frequency is simply the difference between successive resonant frequencies. If it is fixed at one end, the fundamental frequency is one-half the difference between successive resonant frequencies. We will take the ratios of the resonant frequencies to determine which case we have.

Solve:

Determine an *algebraic* expression for the ratio of two successive resonant frequencies (f_{n+1}/f_n) if the string is fixed at both ends. The only variable you should have is n, the mode number.	
Do the same thing if the string is fixed at only one end. Remember that only the odd harmonics are present in this case	
Find the ratio of the high to middle and middle to low frequencies given. Reduce these to a ratio of integers, rather than a decimal value. Do these ratios match the pattern for a string fixed at both ends or a string with one free end?	

From your result, determine the fundamental frequency.	
	$f_1 = 91.0$ Hz

Check: You should have been able to determine that the pattern of ratios matches the pattern for a string fixed at both ends.

Taking It Further: What would the ratios look like (give examples) in order to match the pattern for a string with one free end?

Try It Yourself #4

Two identical strings on a piano are tuned to concert A (440 Hz). Each is under a tension of 1300 N. Over the course of time, one string loosens to the point that, when you strike the two strings simultaneously, you hear beats every 1.10 s. By how much has the tension in the loose string decreased?

Picture: The beat frequency is the same as the difference in the frequency of the two strings. This decrease in frequency comes from a reduction in the speed of the wave on the string, from which you can find the reduction of the tension of the loose string.

Solve:

Find the new frequency from the original frequency and the beat frequency. Because the tension has decreased, the frequency of the loose string decreases, not increases.	
To find the tension in the loose string, take a ratio of the two frequencies. Relate this ratio of frequencies to the ratio of the wave speeds. Remember that the wave speed is a function of the tension and the linear mass density. Since the linear mass density is the same for both strings, the ratio of frequencies can be reduced to depend only on the tensions in the two strings. Use this final result to find the new tension in the loosened string from the original tension and the ratio of frequencies.	

The change in tension can now be calculated. If you carry extra significant figures throughout this calculation, you can get a reasonably precise result in the *change* of the tension. These extra significant figures are required because the change is relatively small.	$\Delta F_T = 5.37$ N

Check: The change in frequency is pretty small, so we don't expect a large change in the tension.

Taking It Further: Will a stringed instrument tend to go sharp (increase in resonance frequency) or flat (decrease in resonant frequency) over time? Why?

16.3   Additional Topics*

In a Nutshell

For a vibrating string of finite length, the motion of the string can always be treated as a superposition of standing waves. The wave function is a sum of the standing wave functions, with each harmonic having a different amplitude. In fact, when a string is plucked, it always vibrates with a combination of harmonic frequencies, not just the fundamental.

When an air column, such as the air in a clarinet or trombone, vibrates, it usually vibrates not as a single standing wave but as the superposition of two or more standing waves. A plot of the pressure versus time for the sound produced is called a **waveform**. Waveforms can be analyzed to determine the amplitudes, frequencies, and phase constants of the harmonics present. This analysis is called **harmonic analysis** or **Fourier analysis**.   The inverse of harmonic analysis—the construction of the original periodic waveform from its harmonic components—is called **harmonic synthesis**.

A periodic (repetitive) waveform, like the sound of a sustained note on a clarinet, is a superposition of harmonic waves whose frequency distribution is discrete (not continuous). A waveform that is not periodic, such as the sound from a firecracker, can be considered a superposition of many harmonic waves whose frequency distribution is continuous. Such a superposition of harmonic waves, which produce a pulse, is called a **wave packet**. For a wave packet, the range of angular frequencies $\Delta\omega$ and the time interval Δt for the packet to pass any point are intricately related.

If a wave packet is to sustain its size and shape, the speed of all the harmonic waves that make up the packet must be the same. This will occur only in a nondispersive medium, that is, a medium in which the speed of a harmonic wave is independent of its wavelength or frequency. In a **dispersive medium**, the speed of harmonic waves varies with frequency. The average speed of the individual harmonic waves that make up a wave packet is called the **phase velocity**, and the velocity of the packet itself is called the **group velocity**.

---

*Optional material.

Fundamental Equations

Temporal wave packet relation	$\Delta\omega\,\Delta t \sim 1$
Spatial wave packet relation	$\Delta k\,\Delta x \sim 1$

Important Derived Results

Superposition of standing wave functions	$y(x,t) = \sum_{n} A_n \sin(k_n x)\cos(\omega_n t + \delta_n)$

Common Pitfalls

5. TRUE or FALSE: A periodic waveform can be synthesized by the superposition of its harmonic components.

6. When we play chords on the piano, several notes (frequencies) are sounded at once. Why don't we hear beats?

QUIZ

1. TRUE or FALSE: The fundamental frequencies of a standing wave in a tube with one end open and in a tube approximately twice as long with both ends open, are the same.

2. TRUE or FALSE: If two wave pulses traveling in the opposite directions along the same string meet, they reflect off each other.

3. The fundamental frequency of an air column in a pipe closed at both ends is 240 Hz; the fundamental frequency is 117.5 Hz if the pipe is open at one end. Why is the ratio of the frequencies approximately 2:1? Why is it not exactly 2:1? What would you expect to be the fundamental frequency of the pipe if it is open at both ends?

4. What happens to the frequency of a clarinet if the clarinetist breathes out pure helium?

5. How can the phenomenon of beats be used to tune musical instruments?

6. The fundamental frequency of a violin string 30.0 cm long is sounded next to the open end of a closed organ pipe 41.0 cm long. The strongest standing sound wave in the pipe occurs when the string is vibrating at the pipe's fundamental frequency. This happens when the tension in the string is 220.0 N. What is the mass of the string? Assume the speed of sound in air is 340 m/s.

7. (a) A man is sitting in a room directly between two loudspeakers 5.00 m apart that are vibrating in phase. He is 1.80 m from the nearer speaker, at point A in the figure. If the lowest frequency at which he observes maximum destructive interference is 122 Hz, what is the speed of sound in air? (b) If, instead, he listens from point B, what is the lowest frequency at which he will observe destructive interference?

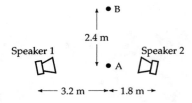

Chapter 17

Temperature and Kinetic Theory of Gases

17.1 Thermal Equilibrium and Temperature

In a Nutshell

A physical property that changes with temperature is called a **thermometric** property. Examples include the volume of a solid and the pressure of a gas whose volume is kept constant, both of which increase when heated.

When two objects are in **thermal contact**, energy in the form of heat flows from the warmer object to the cooler object. This normally results in the warmer object cooling and the cooler object warming. As this process continues, both the rate of cooling of the warmer object and the rate of warming of the cooler object become less and less. When each of the objects is neither warming nor cooling, the objects have reached a steady-state situation called **thermal equilibrium**.

The **zeroth law of thermodynamics** states that if two objects are each in thermal equilibrium with a third, then all three are in thermal equilibrium with one another.

Two objects are defined as having the same **temperature** if they are in thermal equilibrium with each other. The zeroth law, as we will see, enables us to define a temperature scale.

A common thermometer is based on the thermal expansion of a liquid sealed in a glass envelope. For such a thermometer, a temperature scale is established by assigning values at which the thermometer is in thermal equilibrium with reproducible states of some system, such as water at its ice (freezing) and steam (boiling) points. The interval between these points is then divided into equal increments called **degrees**.

For the **Celsius scale**, a value of zero degrees Celsius (0°C) is assigned to the ice point of water, and a value of 100 degrees Celsius (100°C) is assigned to its steam point at atmospheric pressure. A Celsius thermometer is constructed by selecting a convenient thermometric property, measuring that property at the ice and steam points, and dividing the difference between these measured values into 100 equally spaced increments.

In the Fahrenheit scale, still commonly used in the United States, a value of 32 degrees Fahrenheit (32°F) is assigned to the ice point of water, and 212 degrees Fahrenheit (212°F) is assigned to the steam point at atmospheric pressure.

Physical Quantities and Their Units

Temperature   units of °C or °F

Fundamental Equations

Temperature in °C
$$t_{\mathrm{C}} = \frac{L_t - L_0}{L_{100} - L_0} \times 100°$$

Important Derived Results

Fahrenheit–Celsius conversion $t_C = \dfrac{5}{9}\left(t_F - 32°\right)$ or $t_F = \dfrac{9}{5}t_C + 32°$

Common Pitfalls

> ➤ Be careful not to confuse a specific value of temperature with a change in temperature or a temperature range. When dealing with Celsius or Fahrenheit degrees, °C or °F should be used to denote a specific temperature (e.g., 30°C) and C° or F° should be used to denote either a change in temperature or a temperature range (e.g., a 10-C° difference in temperature). Unfortunately, many authors are not consistent in this notation, so you need to rely on the context of the values provided.

1. TRUE or FALSE: If both object A and object B are in thermal equilibrium with object C, then object A must be in thermal equilibrium with object B.

2. Do all Celsius thermometers have to agree that normal body temperature is about 37°C? Why or why not?

17.2   Gas Thermometers and the Absolute Temperature Scale

In a Nutshell

A drawback to most thermometers is that the temperature readings of different kinds of thermometers do not agree, except at the defining temperatures, because different thermometric properties do not change with temperature in the same way. There is one class of thermometers, however, in which temperatures agree over a wide range of values. These are the **constant-volume gas thermometers**, for which pressure is the thermometric property. If any gas at low pressure is confined to a constant volume, its pressure will increase linearly with increasing temperature over a wide range of temperatures.

For gases at very low densities, all constant-volume gas thermometers give the same value of the temperature, independent of the particular gas used. When the pressure in a constant-volume gas thermometer is extrapolated to zero, it approaches zero as the temperature approaches −273.15°C. This temperature is known as **absolute zero**.

The ice and steam points of water are less easy to reproduce precisely than the **triple point** of water, which is the single state at which ice, water, and steam can coexist in equilibrium. This state occurs at a pressure of 4.58 mmHg and a temperature of 0.01°C. The **ideal-gas temperature scale** is defined by making the triple-point temperature 273.16 kelvins (K).

The ideal-gas temperature scale is identical with another scale, the **absolute temperature scale** (also called the **Kelvin scale**), which is defined independently of the properties of any substance. The symbol T is used when referring to the absolute temperature. To convert from degrees Celsius to kelvins, we simply add 273.15.

Physical Quantities and Their Units

Temperature   units of °C or °F or kelvins (K)

Important Derived Results

Celsius scale, gas thermometer	$t_C = \dfrac{P_t - P_0}{P_{100} - P_0} \times 100°C$
Constant-volume ideal-gas temperature scale	$T = \dfrac{273.15 \text{ K}}{P_3} P$
Celsius–kelvin conversion	$T = t_C + 273.15$

Common Pitfalls

> Many of the equations in physics that include a temperature factor are valid only when temperature is expressed in kelvins. Be sure you know for which equations this is true. Of course, if you are dealing with a temperature difference, a kelvin and a Celsius degree are equal.

> Take note that the SI temperature unit, the kelvin, is not a degree and is not accompanied by a degree symbol.

3. TRUE or FALSE: All low-pressure, constant-volume gas thermometers agree that normal body temperature is about 37°C.

4. In what way is the Celsius scale more convenient than the Kelvin scale for everyday use? In what way is the Kelvin scale more suitable than the Celsius scale for scientific use?

Try It Yourself #1

A certain constant-volume gas thermometer reads a pressure of 88.0 torr at the temperature of the triple point of water. (a) What pressure will the thermometer read at a temperature of 310 K? (b) What is the temperature when the thermometer reads a pressure of 70 torr?

Picture: Use the expression for the constant-volume ideal-gas temperature thermometer.

Solve:

Temperature is proportional to pressure in a constant-volume gas thermometer, and we know the pressure and temperature of the triple point of water. Solve for the pressure at 310 K.	$P_{310} = 99.9$ torr
The same concept used in part (a) applies to part (b). Now we know the new pressure, so solve for the new temperature.	$T_{70 \text{ torr}} = 217$ K

Check: At constant volume, lower pressures should correspond to lower temperatures, which they do.

Taking It Further: If a different thermometric property were used, would temperature necessarily vary linearly with changes in that property?

Try It Yourself #2

On a morning when the thermometer reads 52.0°F, you check the pressure in your bicycle tires using your pressure gauge, and the gauge reading is 75.0 lb/in.^{2} Later in the day, after you have ridden several miles on hot pavement, the temperature of the air in the tires reaches 125°F. Assuming that the volume of the tires hasn't changed, what is the pressure in them now?

Picture: This is a constant-volume problem, so the absolute pressure is proportional to the absolute temperature. Because you will be dealing with ratios of pressure, you will not have to convert pressure to units of torr.

Solve:

Convert the temperatures to kelvins. You may have to convert to degrees Celsius as an intermediate stage.	
Convert the pressure readings to absolute pressure. Your tire gauge measures the pressure relative to atmospheric pressure, so you should add the value of atmospheric pressure to each reading.	
At constant volume, pressure is proportional to the temperature. Use this fact to find the new pressure.	$P = 103$ lb/in.^{2} absolute $= 88.0$ lb/in.^{2} gauge

Check: The pressure should go up at higher temperatures.

Taking It Further: Why are you instructed to measure the pressure in car tires only after the car has been sitting idle for an extended period of time?

17.3   The Ideal-Gas Law

In a Nutshell

Experimentally, when a confined gas kept at low pressure P and at a constant temperature T is either compressed or allowed to expand, the product of the pressure of the gas and its volume remains constant. This observation is known as **Boyle's law.**

At low pressures the absolute temperature of a gas kept at a constant volume V is proportional to its pressure. When this observation is combined with Boyle's law, we arrive at the **ideal-gas law**, which states that the pressure times the volume is proportional to the temperature of the gas multiplied by the number of molecules N in the gas. The proportionality constant is called **Boltzmann's constant** $k = 1.38 \times 10^{-23}$ J/K.

An amount of gas is often expressed in moles. A **mole** (mol) of any substance is the amount of that substance that contains **Avogadro's number** $N_A = 6.022 \times 10^{23}$ of atoms or molecules. N_A is defined as the number of carbon atoms in 12 grams of ^{12}C.

The **universal gas constant** $R = N_A k$, can be use to re-express the ideal-gas law in terms of the number of moles of the gas instead of the number of molecules of the gas.

Standard temperature and pressure (STP) refers to a temperature of 0°C and a pressure of 1 atm.

The ideal-gas law is said to be an **equation of state**. By knowing any two of the three quantities pressure, volume, and temperature, you can find the third. These quantities define the macroscopic thermodynamic properties (state) of the gas.

When two or more gases are present in a container, we consider each gas to fill the entire volume. Each gas exerts a **partial pressure** on the walls of the container which is the same pressure it would exert if only that quantity of that gas were present. The **law of partial pressures** states that the total pressure in the container is equal to the sum of the partial pressures.

Solving Dilute-Gas Problems

Picture: A dilute gas is one for which the ideal-gas model gives sufficiently accurate results. The variables are pressure, volume, temperature, mass, and the amount of substance (number of moles).

Solve:

1. Apply the ideal-gas law, $PV = nRT$, to each dilute gas. Be sure to use the absolute temperature and pressure.

2. For a mixture of dilute gases, the ideal-gas law applies to each gas in the mixture, the volume of each gas in the mixture is the volume of the container, and the pressure of each gas is the partial pressure of that gas. The pressure of the mixture is the sum of the partial pressures of the constituent gases.

3. Additional useful relations are $R = N_A k$, $N = n N_A$, and $m = nM$, where k is the Boltzmann constant, N is the number of molecules, m is the mass of the gas, and M is the molar mass.

4. Solve for the desired quantity.

Check: The pressure, volume, and temperature can never be negative.

Physical Quantities and Their Units

Boltzmann's constant $k = 1.38 \times 10^{-23}$ J/K $= 8.617 \times 10^{-5}$ eV/K

Avogadro's number $N_A = 6.022 \times 10^{23}$ molecules/mol

Universal gas constant $R = N_A k = 8.314$ J/(mol $\cdot$ K) $= 0.08206$ L $\cdot$ atm/(mol $\cdot$ K)

Fundamental Equations

Ideal-gas law $PV = NkT = nRT$

Common Pitfalls

➢ In applying the ideal-gas law, be sure that all quantities, including the gas constant R, are in consistent units. Trying to work with pressure in torr, volume in cubic meters, and R in $L \cdot atm/(mol \cdot K)$ will only give you a headache. A good rule is to use only SI units.

➢ If a gas law such as Boyle's law $(P_1 V_1 = P_2 V_2)$ is used for proportional calculations, you can use any units you like, but be sure to use absolute pressure, not gauge pressure. Of course, the units have to be the same on both sides of the equation.

5. TRUE or FALSE: The molar mass of a sample of a gas is the mass of one mole of the gas.

6. What is the mass of a mole of carbon monoxide gas (molecular formula CO)? How many molecules are there in one mole?

Try It Yourself #3

A high-vacuum pump reduces the pressure in a container to 10^{-10} torr. If the temperature is 20°C, how many gas molecules per cubic centimeter are there in the container?

Picture: To use the ideal-gas law, you will have to convert the values for P, V, T, and the gas constant R to compatible units. Units conversions play an important role in this sort of problem.

Solve:

Convert P, V, T, and the gas constant R to compatible units.	
Apply the ideal-gas law to solve for the molar volume.	

Use the fact that there are Avogadro's number of molecules in 1 mol of a gas to find the total number of molecules per cubic centimeter.	
	$N = 3.29 \times 10^6$ molecules/cm^3

Check: Intergalactic space has roughly 25 atoms/cm^3, so in comparison this seems reasonable.

Taking It Further: If you place this container in a liquid nitrogen bath at a temperature of 77 K, what happens to the pressure in the container? To the number of atoms?

Try It Yourself #4

The mass of a certain gas sample that occupies 3.00 L at 20°C and 10.0 atm pressure is found to be 55.0 g. What is this gas?

Picture: Apply the ideal-gas law to find the molar mass of the gas. Use what you remember from chemistry to determine the kind of gas. You may need a periodic table.

Solve:

Arrange the ideal-gas law to find an *algebraic* expression for the number of moles of the gas.	
Use the expression above and the total mass of the gas to find the molar mass of the molecule.	
Determine what kinds of gaseous molecules might have this molar mass. Some possibilities are listed. Can you find others?	
	Carbon dioxide (CO$_2$), nitrous oxide (N$_2$O), propane (C$_3$H$_8$)

Check: These are all reasonable.

Taking It Further: Is this a good way to determine the composition of a material? Why or why not?

17.4 The Kinetic Theory of Gases

In a Nutshell

An ideal gas is one in which the molecules are separated, on the average, by distances that are large compared with their diameters, and they exert no forces on each other except when they collide. The properties of an ideal gas can be understood in terms of a simple kinetic model. We can picture the gas as consisting of noninteracting molecules in motion and the pressure as the result of the collisions of the gas molecules with the walls of their container. The impulse-momentum theorem can be used to relate the pressure to the kinetic energy of the molecules. Combining this with the ideal-gas law yields the following result: $K_{av} = (\frac{1}{2}mv^2)_{av} = \frac{3}{2}kT$. Thus the absolute temperature of the gas is a measure of the **average translational kinetic energy** of its molecules.

The **total translational kinetic energy** K of n moles of a gas containing N molecules is $K = \frac{3}{2}NkT = \frac{3}{2}nRT$.

From these results we can estimate the **root mean square** speeds of the molecules in a gas: $v_{rms} = \sqrt{(v^2)_{av}} = \sqrt{3kT/m}$.

The **equipartition theorem** states that when a substance is in equilibrium, there is an average energy of $\frac{1}{2}kT$ per molecule or $\frac{1}{2}RT$ per mole associated with each degree of freedom.

The average distance λ a gas molecule travels between collisions is called its **mean free path**. For an ideal gas, the mean free path is given by $\lambda = 1/(\sqrt{2}n_V\pi d^2)$, where n_V is the number of molecules per unit volume and d is the molecular diameter. If v_{av} is the average speed, the average distance traveled between collisions is $\lambda = v_{av}\tau$, where τ is the mean time between collisions.

The details of molecular motions can be connected to the macroscopic thermodynamic parameters of pressure, volume, and temperature with statistical mechanics. The **Maxwell-Boltzmann speed distribution function** gives the probability that a given molecule will have a certain speed. The **most probable speed** is given by $v_{max} = \sqrt{2kT/m}$.

The Maxwell-Boltzmann speed distribution can also be written as a translational-kinetic-energy distribution.

Important Derived Results

Average translational kinetic energy $\qquad\qquad K_{av} = \left(\frac{1}{2}mv^2\right)_{av} = \frac{3}{2}kT$

Total translational kinetic energy $\qquad\qquad K = N\left(\frac{1}{2}mv^2\right)_{av} = \frac{3}{2}NkT = \frac{3}{2}nN_AkT = \frac{3}{2}nRT$

Root mean square speed	$v_{\text{rms}} = \sqrt{(v^2)_{\text{av}}} = \sqrt{\dfrac{3kT}{m}} = \sqrt{\dfrac{3RT}{N_A m}} = \sqrt{\dfrac{3RT}{M}}$
Mean free path	$\lambda = \dfrac{1}{\sqrt{2}n_V \pi d^2} = v_{\text{av}}\tau$
Maxwell-Boltzmann speed distribution function	$f(v) = \dfrac{4}{\sqrt{\pi}}\left(\dfrac{m}{2kT}\right)^{3/2} v^2 e^{-mv^2/(2kT)}$
Most probable speed of a gas molecule	$v_{\max} = \sqrt{2kT/m}$
Maxwell-Boltzmann energy distribution function	$F(E) = \dfrac{2}{\sqrt{\pi}}\left(\dfrac{1}{kT}\right)^{3/2} E^{1/2} e^{-E/(kT)}$

Common Pitfalls

> At a given temperature, molecules of all gases have the same average translational kinetic energy. Their average and root mean square speeds will not be the same, however, because the molecular mass of different gases are not the same.

7. TRUE or FALSE: Because gas molecules are so far apart in our atmosphere, and so small, they typically travel several meters before colliding with another gas molecule.

8. Is it possible for the total average kinetic energy of a molecule to be greater than $\frac{3}{2}kT$? Why or why not?

Try It Yourself #5

Naturally occurring uranium has a rare isotope with an atomic mass of 235 and a common isotope with an atomic mass of 238. Uranium reacts with fluorine (atomic weight 19) to form the gas uranium hexafluoride (UF_6). What is the ratio of the rms speed of the UF_6 gas molecules containing atoms of uranium 238 to the rms speed of the gas molecules containing atoms of uranium 235?

Picture: Both types of gas molecules should have the same average translational kinetic energies because they are at the same temperature. You can get atomic masses from Appendix C in the text.

Solve:

Find the masses of uranium 235, uranium 238, and fluorine, as well as the mass of uranium hexafluoride with either species of uranium.	

Using the expression for the average translational kinetic energy, find an *algebraic* expression for the average speed of each species of UF_6.	
Find an *algebraic* expression for the rms speed of each species of the gas.	
Find the ratio. Remember that the average kinetic energy for both species should be the same.	$\dfrac{v_{rms\ 238}}{v_{rms\ 235}} = 0.996$

Check: Since the gas with ^{238}U is more massive, it requires less speed for the same kinetic energy.

Taking It Further: Could this speed difference be useful in separating out the gas made from ^{235}U? Why or why not?

Try It Yourself #6

The surface of the Sun consists primarily of monatomic hydrogen gas at a temperature of 6000 K. Compare the rms speed of a hydrogen atom at the surface of the Sun with the escape speed from the Sun's surface. The Sun's radius is 6.96×10^8 m and its mass is 1.99×10^{30} kg.

Picture: Find the rms speed from the average kinetic energy of hydrogen. You learned how to calculate escape speeds while studying Chapter 11.

Solve:

Find an *algebraic* expression for the escape velocity of the Sun. You may want to refer to Chapter 11.	
Find an *algebraic* expression for the rms speed of monatomic hydrogen in terms of temperature, the universal gas constant, and the molar mass of hydrogen.	
Find the ratio of these two speeds.	$$\frac{v_{\text{rms}}}{v_e} = 0.02$$

Check: Make sure your ratio is unitless.

Taking It Further: Why doesn't the hydrogen just drift off into space?

QUIZ

1. TRUE or FALSE: All Celsius thermometers must agree that the steam point of water is 100°C.

2. TRUE or FALSE: All molecules of the same gas at the same temperature and pressure have the same speed.

3. What change must be made in the temperature of an ideal gas to halve the rms speed of its molecules?

4. What properties should a physical system possess in order to serve as a good thermometer?

5. Two different ideal gases are at the same temperature. How do the rms speeds of their molecules compare? What is the average translational kinetic energy of their molecules?

6. Like the molecules of all monatomic gases, the molecules of neon have no rotational or vibrational kinetic energy, only translational kinetic energy. (a) What is the total kinetic energy of the molecules in 1.00 L of neon gas (atomic mass 22.0) at 1.00 atm pressure? (b) If the gas is expanded at constant temperature to a volume of 2.00 L, by how much does the total kinetic energy change?

7. A rigid, high-pressure gas cylinder has a mass of 21.22 kg when empty. Its interior volume is 1.33 L. Its mass is 21.61 kg after it is filled with equal masses of nitrogen gas (N_2) and oxygen (O_2) at room temperature (20°C). What is the pressure in the filled cylinder?

Chapter 18

Heat and the First Law of Thermodynamics

18.1   Heat Capacity and Specific Heat

In a Nutshell

Heat Q is energy transferred from a warmer object to a cooler object. Once that energy is transferred, it is no longer identified as heat. Instead, it is part of the **internal energy** of the cooler object. The internal energy of an object is its total energy in the center-of-mass reference frame of the object.

The **heat capacity** C is the change in internal energy required to increase the temperature of a sample by one degree. The **specific heat capacity** or **specific heat** c is the heat capacity per unit mass. The **molar specific heat** c' is the heat capacity per mole. Specific heat and molar specific heat are material dependent. Table 18-1 on page 593 of the text lists specific heats and molar specific heats for some materials.

The **calorie** was originally defined as the amount of heat needed to increase the temperature of 1 gram of water 1 Celsius degree. Since we now know heat is simply another form of energy, we define the calorie in terms of the joule: 1 cal = 4.184 J. The U.S. customary unit of heat is the **Btu** or British thermal unit, which was defined as the amount of energy required to increase the temperature of 1 pound of water by 1°F. 1 Btu = 252 cal = 1.054 kJ.

A **calorimeter** is a device used to measure the specific heat of an object. The study of heat transfer is known as **calorimetry**.

Physical Quantities and Their Units

Calorie   1 cal = 4.184 J

British thermal unit   1 Btu = 252 cal = 1.054 kJ

Specific heat of water $c_{\text{water}} = 1 \text{ cal}/(\text{g} \cdot \text{K}) = 1 \text{ Btu}/(\text{lb} \cdot {}^\circ \text{F})$

Important Derived Results

Heat capacity $$C = \frac{Q}{\Delta T} = \frac{\Delta E_{\text{int}}}{\Delta T}$$

Specific heat capacity $$c = \frac{C}{m}$$

Molar specific heat $$c' = \frac{C}{n} = \frac{mc}{n} = Mc$$

Common Pitfalls

> For a solid or a liquid, the heat capacity specified in a table is almost always the heat capacity at a pressure of 1 atmosphere. The heat capacity at constant volume is slightly less than this, but the difference is so small it is usually negligible. This is because, for a given temperature increase, the work done by the solid or liquid in pushing back the atmosphere is small compared to the increase in its internal energy.

> The internal energy of an ideal gas depends only on its temperature. For other systems the internal energy depends on additional conditions, such as the density of the material.

> Be careful of the signs when working thermodynamic problems. You need to keep straight which objects increase in temperature and which objects decrease in temperature.

1. TRUE or FALSE: The only way in which the internal energy of a system can increase is if heat is added to the system.

2. Can the temperature of 2 L of water be increased from 20°C to 30°C without any work being done by the water? Explain.

Try It Yourself #1

A piece of iron (specific heat 0.431 kJ/(kg·K)) of mass 80.0 g at a temperature of 98.0°C is dropped into an insulated vessel containing 120 g of water at 20.0°C. At what final temperature does the system come to equilibrium?

Picture: Assume the heat losses to the container and the surroundings to be negligible. Then the heat lost by the iron is equal to that gained by the water. The heat loss (or gain) is equal to the mass times the specific heat times the temperature change.

Solve:

Convert temperatures to kelvins. Because we are dealing with only temperature differences, in this problem the conversion is not strictly necessary, but it is good practice.	
Convert masses to kilograms.	

Write *algebraic* expressions for the heat gained by the water and the heat lost by the iron. These expressions should include the initial and final temperatures of the objects. Make sure you use meaningful subscripts so you can keep track of which material you are describing.	
Equate the heat lost by the iron to the heat gained by the water. From this expression, find an *algebraic* expression for the final temperature. Both the water and the iron will have the same final temperature.	
Substitute values into the expression above and find the final temperature.	$T_f = 298 \text{ K} = 25°\text{C}$

Check: Our intuition tells us the water should increase in temperature and the iron should decrease in temperature. This is exactly what we see.

Taking It Further: Why does the temperature of the water increase so much less than the iron cools?

Try It Yourself #2

Imagine that you want to take a warm bath, but there's no hot water. You draw 40.0 kg of tap water at 18°C in the bathtub and heat water on your stove to warm the bath water up. If you heat the water to 100°C in a 2.00-L saucepan, how many panfuls must you add to the bath to raise its temperature to 40°C?

Picture: If the heat losses to the bathtub and the surroundings are negligible, the heat lost by the heated water is equal to that gained by the cold bath water. The heat lost (or gained) is equal to the mass times the specific heat times the temperature change.

Solve:

Find *algebraic* expressions for the heat gained by the tub water and the heat lost by the stove water. These expressions should include the initial and final temperatures and the mass of both bodies of water. Make sure to use meaningful subscripts so you can keep track of which mass is which.	
Equate the heat gained by the tub water to the heat lost by the stove water to determine the mass of stove water required.	
From the density of water, determine how many liters of stove water are required.	
Now determine how many pans of water this will take.	7.33 pans

Check: This seems like a reasonable number.

Taking It Further: Will the actual number of pans of hot water required be greater than, less than, or equal to the number calculated? Why?

18.2 Change of Phase and Latent Heat

In a Nutshell

Under certain conditions a substance will remain at the same temperature while energy is being transferred to or from it in the form of heat. This happens when the substance undergoes a **phase change** such as **fusion** (a change from the liquid phase to the solid phase). The other common phase changes are **vaporization** (a change from a liquid to a vapor or gas), **sublimation** (a change from a solid directly into a gas), and **melting** (a change from a solid phase to the liquid phase.)

The **latent heat of fusion** L_f is the heat per unit mass required to melt a substance, or the heat given up by the substance as it solidifies.

The **latent heat of vaporization** L_v is the heat per unit mass required to change the phase of a substance from a liquid to a gas, or the heat given up by the substance as its phase changes from a gas to a liquid.

There are many other kinds of phase changes, as well, such as the change of a solid from one crystalline form to another, in which the internal configuration of the atoms or molecules in the solid change. An example is the conversion of carbon graphite (a solid) to diamond (another solid) while under very high pressure.

Physical Quantities and Their Units

Latent heat of fusion for water	$L_f = 333.5 \text{ kJ/kg} = 79.7 \text{ kcal/kg}$
Latent heat of vaporization for water	$L_v = 2.26 \text{ MJ/kg} = 540 \text{ kcal/kg}$

Important Derived Results

Latent heat of fusion	$L_f = \dfrac{Q_f}{m}$
Latent heat of vaporization	$L_v = \dfrac{Q_v}{m}$

Common Pitfalls

> If when an object is heated or cooled the range of temperatures passes through the boiling or melting point of the material, remember to account for the latent heat of fusion or vaporization.

3. TRUE or FALSE: Removing heat from a system always decreases its temperature.

4. Can a system absorb heat without its temperature increasing? Can a system absorb heat without its internal energy increasing? Explain.

Try It Yourself #3

A copper bar of mass 2.50 kg at an initial temperature of 66.0°C is dropped into an insulated vessel containing 400 g of water and 70.0 g of ice at 0.00°C. The specific heat of copper can be found in Table 18-1 on page 593 of the text. At what final temperature does the system come to thermal equilibrium?

Picture: Assume the heat losses to the container and the surroundings to be negligible. Unlike the previous examples, here we need to consider the effects of the melting ice. The ice must melt before the water temperature increases. As a result, the heat lost by the copper is equal to the latent heat of fusion required to melt the ice plus the heat gained by the resulting total amount of water.

Solve:

Determine the amount of heat required to melt the ice.	
Determine how much the temperature of the copper will drop if it gives up this much heat.	
Find the final temperature of the copper after the ice has melted. Because the copper is still warmer than the water, it will increase the temperature of the water.	
The remaining heat lost by the copper is equal to the heat gained by the *total* mass of water, which must include the mass of the newly melted water. Use this to solve for the final temperature of the water.	$T_\mathrm{f} = 13.8°\mathrm{C}$

Check: The water warmed, and the copper cooled, as expected. The water did not warm as much as the copper cooled because the copper had to first melt the ice, and because the heat capacity of water is so much larger than copper.

Taking It Further: What would it mean if the temperature of the copper in the third step above was less than 0°C?

Try It Yourself #4

An abundance of steam at 100°C is passed into an insulated flask containing 200 g of ice at −25.0°C. Neglecting any heat loss from the flask, what mass of liquid water at 100°C will finally be present in the flask? The specific heats of ice and liquid water can be found in Table 18-1 on page 593 of the text. The latent heats of fusion and vaporization of water can be found in Table 18-2 on page 596 of the text.

Picture: The temperature of the ice will first be raised to 0°C, then the ice will melt, and finally the temperature of the melted water will be raised to 100°C. The heat gained by the ice in this process will be equal to the heat lost by some of the steam as it condenses. The final mass of water at 100°C will be equal to the initial mass of the ice plus the mass of the condensed steam.

Solve:

Determine how much heat is required to raise the temperature of the ice to 0°C.	
Determine how much heat is required to melt the ice.	
Determine how much heat is required to raise the temperature of the water from 0°C to 100°C.	

The total heat gained by the ice in the previous steps is equal to the heat lost by the steam that condenses. Use this to solve for the mass of the condensed steam.	
The total mass of liquid water at 100°C is the sum of the mass of the initial ice and the condensed steam.	$m = 271$ g of water

Check: We end up with a greater mass of water than the initial mass of ice. This is consistent with our understanding that all the ice would melt, and the heat for this process would come from some of the steam condensing.

Taking It Further: How would this problem change if the steam were at 150°C instead of 100°C?

18.3 Joule's Experiment and the First Law of Thermodynamics

In a Nutshell

The **mechanical equivalence of heat** is the fact that 4.184 J of mechanical energy is exactly equivalent to 1 cal of heat.

The **first law of thermodynamics** states that the change in the internal energy of the system equals the heat transfer into the system plus the work done on the system. The heat transferred into the system is Q_{in}. If heat is removed from the system, then Q_{in} is negative. The work done *on* the system is W_{on}. If the system does work on the surroundings, then W_{on} is negative.

Pressure, volume, temperature, and internal energy are state variables—that is, they can describe the system at any point in time. Work and heat, however, are *not* state variables. Rather, they are transfers of energy at some point in time.

Fundamental Equations

First law of thermodynamics	$\Delta E_{int} = Q_{in} + W_{on}$
Differential first law of thermodynamics	$dE = dQ_{in} + dW_{on}$

Common Pitfalls

> ➤ Be very careful with signs in first-law problems, heat-exchange problems, and such. It's easy to get them mixed up. The convention is that Q_{in} is positive when heat flows into the system and W_{on} is positive when work is done by the surroundings on the system.

5. TRUE or FALSE: If positive work is done on a system, then its internal energy must increase.

6. Give an example of a process in which the temperature of a system is increased without any heat input or output.

Try It Yourself #5

Consider a 1.00-mol block of ice as it slowly warms from a temperature of $-10°C$ to $-3°C$. As it warms, it expands 21.0 mm^3. The molar mass of ice is 18.0 g/mol. (a) Calculate the work done by the ice as it expands the 21.0 mm^3 at atmospheric pressure. What does it do this work on? (b) How much heat is transferred to the block during this process? (c) What is the increase in the block's internal energy?

Picture: The work at constant pressure is the product of pressure and the change in volume. Use the values for specific heats and molar heat capacities of solids and liquids in Table 18-1 on page 593 of the text. Use the first law of thermodynamics to find the change in internal energy.

Solve:

Determine the work done by the ice. The work done by the ice on the atmosphere is the opposite of the work done by the atmosphere on the ice. This work is done at constant pressure, and is the product of the pressure and change in volume.	$W_{by\ ice} = 2.1210^{-3}$ J
Heat is transferred to the ice under constant pressure. Determine the amount of heat transferred from the change in temperature and the heat capacity of ice.	$Q_{in} = 258$ J
The total change in the internal energy can be found from the first law of thermodynamics. Because the ice does positive work on the atmosphere, the atmosphere does negative work on the ice.	$\Delta E_{int} = 258$ J

Check: The work done when a solid expands (or contracts) because of a temperature change is typically quite small compared to the heat gained or lost, as this problem demonstrates.

Taking It Further: If the ice were to melt and the temperature increase to 4°C, how would this problem change?

18.4   The Internal Energy of an Ideal Gas

In a Nutshell

At low densities, the total volume of the molecules of a gas, compared with the volume occupied by the gas, is negligible, and the forces the molecules exert on one another during the intervals between collisions are negligible. Gases at densities low enough to fulfill these conditions are called **ideal gases**.

The **average translational kinetic energy** of the molecules of an ideal gas is proportional to the absolute temperature of the gas. If the internal energy E_{int} of a monatomic gas consists only of this translational kinetic energy, then $E_{int} = K = \frac{3}{2}nRT$, where n is the number of moles of a gas and R is the universal gas constant.

To test whether or not the particles of real gases are non-interacting, Joule performed an experiment in which a gas confined in a rigid, thermally insulated container was allowed to expand into a second rigid, thermally insulated container that, prior to the expansion, was evacuated. This process is called a **free expansion**.

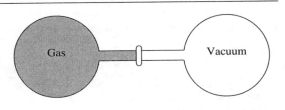

In a free expansion the gas does no work on its surroundings and no heat is transferred to or from the gas, so the internal energy of the gas does not change. Joule found that if a gas initially at low density (with the molecules well separated) underwent a free expansion, any changes in its temperature were too small to observe. However, when a gas initially at high density (with the molecules in close proximity) underwent a free expansion, he observed a slight decrease in its temperature. These results demonstrate that for a gas at low density, the interactions between molecules are negligible; however, for a gas that is at a higher density, the attractive forces between molecules have a small effect.

Common Pitfalls

> ➤ In a free expansion the gas does no work. For any other type of expansion, negative work is done on the gas. As a gas is compressed, positive work is done on the gas.
> ➤ The internal energy of an ideal gas depends only on its temperature. For other systems the internal energy depends on additional conditions, such as the density of the material.

7. TRUE or FALSE: The internal energy of an ideal gas depends only on the temperature of the gas.

8. An ideal gas expands at constant temperature and does work on a piston. Does the internal energy of the gas decrease? If not, where does the energy to do the work come from?

18.5   Work and the *PV* Diagram for a Gas

In a Nutshell

In a **quasi-static process** a gas moves through a series of equilibrium states while the pressure or volume of the gas is slowly changed.

During a quasi-static process the work done on the gas is $W_{\text{on gas}} = -\int_{V_i}^{V_f} P\, dV$. During an expansion the work done on the gas is negative, and during a compression the work done on a gas is positive. The magnitude of the work done is the area under a *PV* curve. The sign depends on the direction of change, so a **directed line** is always used to indicate the direction of the change the gas experiences.

An **isobaric** compression or expansion occurs at constant pressure.

If a gas is heated or cooled at constant volume, we say the process is **isometric**, **isochoric**, or **isovolumetric**. Because the volume remains constant for these processes, the work done on the gas during these transformations is zero.

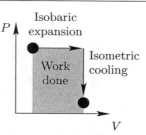

Calculating Work Done by an Ideal Gas During a Constrained Quasi-Static Process
Picture: The increment of work done by a gas is equal to the pressure times the increment of volume—that is, $dW_{\text{by}} = P\, dV$. It follows that $W_{\text{by}} = \int_{V_i}^{V_f} P\, dV$. The constraint dictates how to evaluate this integral.
Solve:

1. If the volume V is constant, then dV equals zero and $W_{\text{by}} = 0$.
2. If the pressure P is constant, then $W_{\text{by}} = P\int_{V_i}^{V_f} dV = P(V_f - V_i)$.
3. If the temperature T is constant, then $P = nRT/V$ and

$$W_{\text{by}} = \int_{V_i}^{V_f} \frac{nRT}{V}\, dV = nRT \int_{V_i}^{V_f} \frac{dV}{V} = nRT \ln \frac{V_f}{V_i}$$

4. If no energy is transferred to or from the gas via heat, then see Section 18-9.
Check: If the volume of the gas increases then W_{by} must be positive, and vice versa.

Important Derived Results

Liter-atmospheres $\qquad\qquad$ $1\ \text{L} \cdot \text{atm} = (10^{-3}\ \text{m}^3)(1.013 \cdot 10^5\ \text{Pa}) = 101.3\ \text{J}$

Work done on a gas $\qquad\qquad$ $W_{\text{on gas}} = -\int_{V_i}^{V_f} P\, dV = \text{negative area under } PV \text{ curve}$

Common Pitfalls

> ➢ In doing first-law-of-thermodynamics problems, don't assume that a heat transfer and a temperature change always go together. A quasi-static adiabatic expansion is an example of a temperature change without heat transfer. An isothermal expansion is an example of a heat transfer without a temperature change. There can be a temperature change without any heat transfer, and there can be a heat transfer without any temperature change.

> ➢ When an ideal gas undergoes some process and thus changes state, the ideal-gas law may be used to relate initial and final states. However, this equation does not specify whether heat

flowed into or from the gas in the process. The amount of heat exchanged by the gas depends on how the process was carried out.

> Pay careful attention to the words "by" and "on". These words tell you the direction of energy transfer, from which you can determine the appropriate signs for the phenomena described in a problem.

9. TRUE or FALSE: In an isometric compression, no work is done on a gas.

10. An ideal gas expands slowly to twice its initial volume (a) at constant pressure and (b) at constant temperature. In which case does it do more work on its surroundings? Why?

Try It Yourself #6

In the *PV* diagram for a monatomic ideal gas shown in the figure, path A is an isothermal expansion and path B is an expansion at a constant pressure followed by cooling at constant volume. For each process, calculate the heat input, the work done, and the change in internal energy for 1 mol of the gas.

Picture: We will use the problem-solving strategy outlined above.

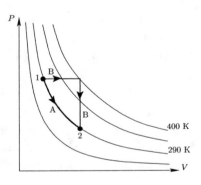

Solve:

For path A, use the expression for a quasi-static isothermal expansion to find an *algebraic* expression for the work done on the gas in terms of the initial and final volumes.	
Find the ratio V_i/V_f by realizing that the ratio of the volumes on the constant pressure segment of path B is also V_i/V_f, and that for an expansion at constant pressure, the volume is directly proportional to the temperature.	
Solve for the work done on the gas via path A on the *PV* curve.	$W_{\text{A, on gas}} = -775$ J

Find the change in the internal energy of the gas for path A. Remember that path A is an isothermal process.	
	$\Delta E_{\mathrm{int}} = 0$
Use the first law of thermodynamics to calculate the heat added to the gas.	
	$Q_{\mathrm{in}} = 775$ J
For path B, find the work done during the first segment of the path, which is an isobaric expansion. The ideal gas law provides relationships between temperature, pressure, and volume at the two points.	
For path B, find the work done during the second segment of the path, which is isometric cooling.	
Find the total work done on the gas along path B.	
	$W_{\mathrm{B,\ on\ gas}} = -914$ J
Determine the change in the internal energy of the gas for path B. Remember that the initial and final temperatures are the same.	
	$\Delta E_{\mathrm{int}} = 0$

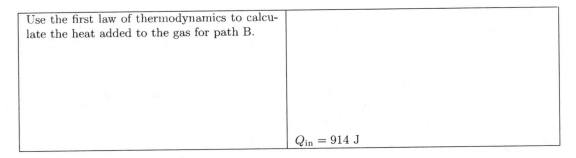

Check: Even though the initial and final temperatures are the same, heat is still added to the system to facilitate the expansion of the gas.

Taking It Further: Assuming you have complete control of P, V, and T, what will be the most efficient way (requires the least heat input) to get from point 1 to point 2 on the PV graph?

18.6   Heat Capacities of Gases

In a Nutshell

If a sample of a material is allowed to expand as it is heated at a constant pressure (as nearly all materials tend to do), more heat input is required for a given temperature increase than would be required for the same temperature increase if the volume of the material were held constant. This is because in expanding, the sample does work by pushing its surroundings back. This loss of energy to the surroundings is compensated by an increased input of heat.

For solids and liquids, which expand only slightly, the additional heat required for a given temperature increase is very small. Thus, the specific heat at constant pressure is approximately equal to the specific heat at constant volume. The specific heat at constant pressure is normally specified because solids and liquids generate enormous pressure if they are not allowed to expand, making it difficult to measure the specific heat at constant volume.

The heat capacity at constant pressure for gases is substantially greater than the heat capacity at constant volume. At constant pressure, gases undergo significant fractional increases in volume, so an appreciable increase of heat is required for a given temperature change. The relation between the heat capacity at constant pressure and the heat capacity at constant volume is $C_P = C_V + nR$.

The heat capacity for a gas at constant volume is $C_V = dE_{int}/dT$. According to the equipartition theorem, there is an energy of $\frac{1}{2}kT$ for each degree of freedom. For a monatomic ideal gas, with three translational degrees of freedom, the total internal energy is $E_{int} = \frac{3}{2}kT$. Diatomic molecules have two additional degrees of freedom for rotation, so their total internal energy is $E_{int} = \frac{5}{2}kT$. Molecules with still more atoms have even more degrees of freedom.

Important Derived Results

Heat capacity for a gas at constant volume $C_V = \dfrac{dE_{int}}{dT}$

Heat capacity for a gas at constant pressure $\qquad$ $C_P = C_V + nR$

Internal energy per mole for monatomic gases $\qquad$ $E_{\text{int}} = \frac{3}{2}RT$

Internal energy per mole for diatomic gases $\qquad$ $E_{\text{int}} = \frac{5}{2}RT$

Common Pitfalls

➤ Remember that the heat capacities C_P and C_V refer to a specific sample of gas and are proportional to the quantity of gas. By contrast the molar heat capacities (heat capacities per mole) c_P and c_V are independent of the total amount of gas.

11. TRUE or FALSE: The heat capacity of a material is its specific heat per unit mass.

12. Consider nitrogen (N_2) and helium (He) gases. For which gas is the internal energy per mole greater at a given temperature? For which gas is the internal energy per gram greater at a given temperature?

Try It Yourself #7

One-half mole of nitrogen gas is heated from room temperature (20°C) and a pressure of 1.00 atm to a final temperature of 120°C. (a) How much heat must be supplied if the volume is kept constant while the gas is heated? (b) How much heat must be supplied if the heating is at constant pressure? (c) By how much is the internal energy changed?

Picture: Remember that nitrogen is a diatomic gas. The molar heat capacities of nitrogen can be found in Table 18-3 on page 607 of the text. For an ideal gas, the change in internal energy depends only upon the change in temperature.

Solve:

For part (a), use the first law of thermodynamics. If the volume remains constant, no work is done on the gas.	
	$Q = 1040$ J

For part (b), the surroundings will do work on the gas.	$Q = 1455$ J
The change in internal energy is due to the change in temperature.	$\Delta E_{\text{int}} = 1040$ J

Check: More heat must be supplied in the constant pressure case, as expected.

Taking It Further: Clearly if we replace the nitrogen gas with one-half mole of a monatomic gas less heat will be required to cause the same temperature change because the molar heat capacity of a monatomic gas is smaller than for the diatomic gasses. What can you say about the required heat if instead of replacing the nitrogen gas molecule for molecule it was replaced gram for gram by a monatomic gas?

Try It Yourself #8

For 195 g of a certain ideal gas, the heat capacity at constant volume is 145 J/K and the heat capacity at constant pressure is 203 J/K. (a) How many moles of the gas are there? (b)What is its molar mass?

Picture: You will need to use the relation between the constant volume and constant pressure heat capacities to find the number of moles.

Solve:

Use the expression relating the heat capacities at constant volume and constant pressure to find the number of moles n.	$n = 6.98$ mol

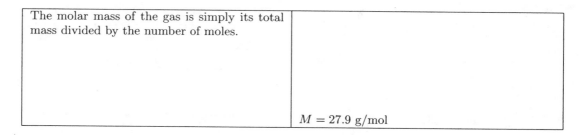

The molar mass of the gas is simply its total mass divided by the number of moles.	
	$M = 27.9$ g/mol

Check: The number of moles and molar mass should by definition be positive quantities, and they are.

Taking It Further: What might this gas be?

18.7 Heat Capacities of Solids

In a Nutshell

The **Dulong-Petit law** states that most solids have molar heat capacities approximately equal to $3R$. The molecules in a solid are fixed in place, but each can vibrate in three dimensions. Solids thus have a total of six degrees of freedom (three for vibrational potential energy and three for vibrational kinetic energy), and the equipartition theorem predicts a heat capacity per mole of $3R$.

Important Derived Results

Dulong-Petit law for solids $\qquad\qquad\qquad c = 3R$

Try It Yourself #9

A 160-g mass of a certain metal at an initial temperature of 88°C is dropped into 140 g of water in an insulated container. The water is initially at 10°C. The system finally comes to equilibrium at 18.4°C. (a) If heat losses to the container and the surroundings are negligible, what is the specific heat of the metal? (b) If the Dulong-Petit law holds, what is the molar mass of the metal?

Picture: If the heat losses to the container and the surroundings are negligible, the heat lost by the metal object is equal to that gained by the water. The heat lost (or gained) is equal to the mass times the specific heat times the temperature change. Find the heat capacity per mole of the metal and by using the Dulong-Petit law, determine the molar mass of the metal, and hence the composition of the metal.

Solve:

Find an *algebraic* expression for the heat gained by the water.	
Find an *algebraic* expression for the heat lost by the metal.	
Equate the heat lost by the metal to the heat gained by the water to solve for the specific heat of the metal.	$c_{\text{metal}} = 0.441 \text{ kJ/(kg·K)}$
The heat capacity per mole is the product of the specific heat and the molar mass M. The heat capacity per mole for a solid is approximately $3R$, according to the Dulong-Petit law. Use this to find the molar mass of the metal.	$M = 56.5 \text{ g/mol}$

Check: This is a reasonable molar mass.

Taking It Further: What kind of metal could this be?

18.8   Failure of the Equipartition Theorem

In a Nutshell

Although the equipartition theorem has been highly successful, it has shortcomings. Ultimately it fails because at the molecular level Newtonian mechanics fails. To predict observed results successfully, the theory of quantum mechanics must be used. In quantum mechanics energy is not continuous, but **quantized**—that is only specific energy levels are allowed.

If the spacing of the allowed energy levels is large compared with kT, energy cannot be transferred by collisions and the classic equipartition theorem is not valid. If the spacing of the levels is much smaller than kT, energy quantization will not be noticed and the equipartition theorem will hold.

18.9   The Quasi-Static Adiabatic Compression of a Gas

In a Nutshell

An **adiabatic process** is one in which there is no net flow of heat into or out of the system.
A **quasi-static process** is one that happens in infinitesimal steps with delays between steps allowing the system to reach equilibrium.
If a gas undergoes a quasi-static adiabatic expansion (or compression), it can be shown that $PV^\gamma = $ constant and $TV^{\gamma-1} = $ constant, where $\gamma = C_P/C_V$ is the ratio of the constant pressure and constant volume heat capacities.
The speed of sound in a medium can be derived from the fact that $PV^\gamma = $ constant for a quasi-static adiabatic process.

Important Derived Results

Quasi-static adiabatic process $\qquad\qquad$ $PV^\gamma = $ constant and $TV^{\gamma-1} = $ constant

Quasi-static adiabatic work $\qquad\qquad$ $W_{\text{on gas, adiabatic}} = C_V \, \Delta T = \dfrac{P_f V_f - P_i V_i}{\gamma - 1}$

Ratio of specific heats $\qquad\qquad$ $\gamma = \dfrac{C_P}{C_V} = \dfrac{c_p}{c_V}$

Common Pitfalls

> ➤ In doing first-law-of-thermodynamics problems, don't assume that a heat transfer and a temperature change always go together. A quasi-static adiabatic expansion is an example of a temperature change without heat transfer. An isothermal expansion is an example of a heat transfer without a temperature change. There can be a temperature change without any heat transfer, and there can be a heat transfer without any temperature change.

13. TRUE or FALSE: In an adiabatic process, the internal energy of the system remains constant.

14. The ratio of molar heat capacities ($\gamma = \frac{c_p}{c_V}$) for a monatomic ideal gas is $\frac{5}{3} = 1.67$; for a diatomic ideal gas it is $\frac{7}{5} = 1.4$. Would you expect the ratio to be higher or lower than 1.4 for an ideal gas having three or more atoms per molecule? Why?

Try It Yourself #10

Two moles of nitrogen, initially at a temperature of 293 K, undergo a quasi-static, adiabatic expansion from a pressure of 5 atm to a pressure of 1 atm. Find the work done on the gas.

Picture: For any process involving an ideal gas, the change in internal energy depends only upon the change in temperature of the gas. For an adiabatic process, the change in internal energy plus the work done by the gas equals zero. For a quasi-static adiabatic expansion of an ideal gas, PV^γ is constant. Use this together with the ideal-gas law to find the temperature of the gas.

Solve:

Use the relationship for an adiabatic process to find an *algebraic* expression for the work done on the gas.	
You must first find the final temperature, using the fact that this is a quasi-static adiabatic expansion (PV^γ = constant), and the ideal-gas law. Your *algebraic* expression will include the initial temperature, the initial and final pressure, and γ.	
Solve for γ before substituting back in for the work done on the gas. Remember that nitrogen is a diatomic gas.	
Finally, substitute all values into the expression found in the first step and solve.	$W = -4490$ J

Check: For the pressure to go down, the volume must increase, so negative work must be done on the system, which is what we found.

Taking It Further: Is the final temperature higher or lower than the initial temperature? Why?

Try It Yourself #11

Ten grams of argon at a pressure of 1.00 atm are placed in a thermally insulated, flexible 4.00-L container. The container and the argon are slowly lowered into the ocean until the volume occupied by the argon is 1.00 L. (a) What is the pressure of the gas after being compressed? (b) How much work is done on the gas as it is compressed from 4 L to 1 L? (c) What is the change in internal energy of the gas?

Picture: Argon is a monatomic gas. The number of moles of argon can be obtained from the given mass and the information in Appendix C of the text. For a quasi-static adiabatic compression of an ideal gas, PV^γ is constant. Use this to find the final pressure of the gas. There is a formula for the work done by the gas in a quasi-static adiabatic expansion or compression. For any adiabatic process, the change in the gas's internal energy plus the work done by the gas equals zero.

Solve:

Find the pressure of the gas after being compressed by using the fact that PV^γ is constant.	$P = 10.1$ atm
Use the equation for the work done on a gas during adiabatic expansion to find the work done.	$W_{\text{on gas}} = 9.12$ L·atm
Use the fact that the change in internal energy plus the work done on the gas is equal to zero to find the change in internal energy.	$\Delta E_{\text{int}} = +9.12$ L·atm

Check: Because the gas is compressed, we expect the pressure to increase, and the work done on the gas to be positive, which is what we found.

Taking It Further: Is the final temperature higher or lower than the initial temperature? Why?

QUIZ

1. TRUE or FALSE: A quasi-static process is one in which no work is done.

2. TRUE or FALSE: In an isothermal process, the temperature of the system remains constant.

3. For solids and liquids, we usually don't distinguish between specific heats at constant volume and specific heats at constant pressure. Why not? If we did want to make this distinction, which one would be larger? Which specific heat are you most likely to find tabulated?

4. Give an example of a process in which the temperature of a system is increased without any heat input or output.

5. You wish to move from a high-volume point on an isotherm to a low-volume point on the same isotherm. Assuming you can control P, V, and T with complete efficiency, what series of steps can you take that would be the absolute most efficient (requires the least net heat input)?

6. To do a spectacular "jumping ring" demonstration you need to cool a ring of aluminum (specific heat 0.900 kJ/(kg·K)) of mass 135 g that is initially at 20°C. You place it in a large container of liquid nitrogen at 77.0 K (its normal boiling point). If the latent heat of vaporization of nitrogen is 199 kJ/kg, what mass of nitrogen is vaporized in cooling the aluminum to 77.0 K?

7. The specific heat of copper is 0.386 kJ/(kg·K). If 180 g of copper at 200°C is dropped into an insulated container containing 280 g of water at 20°C, what is the final equilibrium temperature of the copper and water?

Chapter 19

The Second Law of Thermodynamics

19.1 Heat Engines and the Second Law of Thermodynamics

In a Nutshell

The **Kelvin statement** of the second law of thermodynamics is: No system can absorb heat from a single reservoir and convert it entirely into work without additional net changes in the system or its surroundings.

The **Clausius statement** of the second law of thermodynamics is: A process whose only net result is to absorb heat from a cold reservoir and release the same amount of heat to a hot reservoir is impossible.

The **heat-engine statement** of the second law of thermodynamics is: It is impossible for a heat engine working in a cycle to produce *only the effect* of absorbing heat from a single reservoir and performing an equivalent amount of work.

All statements of the second law of thermodynamics are equivalent.

A **heat engine** is a cyclic device that extracts heat from a high-temperature heat reservoir, converts some of the heat into mechanical energy, and transfers the mechanical energy, as work, to some external agent. In accordance with the first law of thermodynamics, the heat that is not converted into work is rejected to a low-temperature reservoir.

A **heat reservoir** is a system with a heat capacity so large that it can absorb or give off heat with no appreciable change in its temperature. The surrounding atmosphere, lakes, and oceans often act as practical heat reservoirs.

A heat engine absorbs heat Q_h from a high-temperature reservoir, converts some of the heat into work W done by the engine on its surroundings, and rejects the remaining heat Q_c to a low-temperature reservoir. <u>Note</u>: Our sign convention is that W, Q_h, and Q_c represent magnitudes, and as such are always positive. Application of the first law of thermodynamics gives the work done by the heat engine during one cycle as $W = Q_h - Q_c$.

The efficiency ε of the engine is the ratio of the work done by the heat engine to the heat absorbed by the engine from the high-temperature reservoir. In practical engines, the efficiency may be as high as 50% or more, but it is impossible to make a heat engine with an efficiency of 100%, according to the heat-engine statement of the second law of thermodynamics.

Calculating the Work Done by a Heat Engine Operating in a Cycle

Picture: A heat engine absorbs heat from a high-temperature heat reservoir, does work, and releases heat to a low-temperature heat reservoir. The law of conservation of energy informs us that the heat absorbed by the engine per cycle equals the heat released by the engine per cycle plus the work done by the engine per cycle. The efficiency of a heat engine is defined as the ratio of the work done by the engine per cycle to the heat absorbed by the engine per cycle. The working substance—the substance that absorbs heat, such as $H_2 0$ in a steam engine—for the engine is an ideal gas for virtually all calculations in this text.

Solve:

1. For an integral number of cycles, the change in internal energy $\Delta E_{int} = 0$, so $Q_h = W + Q_c$.
2. The efficiency is given by $\varepsilon = W/Q_h$.
3. The work done by the engine during a step in a cycle is given by $W_{step} = \int_{V_i}^{V_f} P \, dV$, where
   $P = nRT/V$.
4. The heat absorbed by the gas during a step is given by $C \, \Delta T$, where C is the heat capacity.

Check: The work done by the engine W must be equal to $Q_h - Q_c$ if the engine completes an integral number of cycles.

Important Derived Results

Work done by a heat engine in one cycle $\qquad\qquad W = Q_h - Q_c$

Efficiency of a heat engine $\qquad\qquad\qquad\qquad \varepsilon = \dfrac{W}{Q_h} = \dfrac{Q_h - Q_c}{Q_h} = 1 - \dfrac{Q_c}{Q_h}$

Common Pitfalls

> The total work done by a heat engine during one cycle is the work done by the working substance during its expansion plus the work done by it during its compression, not just the work during the expansion. (The work done by the engine during the expansion is positive, whereas during the compression the work is negative.)

1. TRUE or FALSE: Work can never be converted completely into internal energy.

2. When we say "engine," we think of something mechanical with moving parts. In such an engine, friction always reduces the engine's efficiency. Why is this?

Try It Yourself #1

A certain engine absorbs 150 J of heat and rejects 88.0 J in each cycle. (a) What is its efficiency? (b) If it runs at 200 cycles/min, what is its power output?

Picture: In any complete cycle, the heat the engine absorbs equals the work it does plus the heat it rejects.

Solve:

Find the amount of work the engine does in one complete cycle.	
The efficiency is the work done by the engine divided by the amount of heat it absorbs.	$\varepsilon = 41\%$
The power is the work done per unit time. In 1 minute, the engine goes through 200 cycles.	$P = 207$ W

Check: This power and efficiency seem reasonable.

Taking It Further: At what rate must we provide energy to the engine?

19.2 Refrigerators and the Second Law of Thermodynamics

In a Nutshell

A **refrigerator** is simply a heat engine run backward. It operates in a cyclic process, in which work W is done by an external agent on the refrigerator, heat Q_c is absorbed from a low-temperature reservoir by the working substance, and heat Q_h is rejected to a high-temperature reservoir.

The **coefficient of performance** (COP) for a refrigerator is a measure of a refrigerator's performance. It is defined as the ratio of the heat absorbed from the low-temperature reservoir to the work done on the refrigerator.

The **refrigerator statement of the second law of thermodynamics** is: It is impossible for a refrigerator working in a cycle to produce no effect other than the transfer of energy as heat from a cold object to a hot object. An ideal refrigerator would require zero work input and would transfer heat from a lower- to a higher-temperature reservoir without any other change to the surroundings. This is impossible according to the second law.

Although appearing to be very different, the heat-engine and refrigerator statements of the second law are equivalent in that if we could make a device that violated one statement, it would also violate the other. This can be seen from simple heat-flow diagrams and is independent of any specific engine design features. See the text for more information.

Important Derived Results

Coefficient of performance for a refrigerator $\text{COP} = \dfrac{Q_c}{W}$

Common Pitfalls

3. TRUE or FALSE: According to the second law of thermodynamics, a refrigerator cannot have a COP of 1 or greater.

4. There are people who try to keep cool on a hot summer day by leaving the refrigerator doors open, but you can't cool your kitchen this way! Why not?

Try It Yourself #2

An electric refrigerator removes 13.0 MJ of heat from its interior for each kilowatt-hour of electric energy used. What is its coefficient of performance?

Picture: The coefficient of performance is the ratio of the heat extracted by the refrigerator to the work done on the refrigerator by an external agent.

Solve:

Convert the work done on the refrigerator (1 kilowatt-hour) to joules.	
Solve for the coefficient of performance (COP).	$\text{COP} = 3.6$

Check: This seems to be a reasonable value for the COP.

Taking It Further: If the coefficient of performance is greater than 1, do we get more energy out than we put in, violating conservation of energy? Why or why not?

Try It Yourself #3

A certain refrigerator has a power rating of 88.0 W. Consider it to be a reversible refrigerator. If the temperature of the room is 26.0°C, how long will the refrigerator take to freeze 2.50 kg of water that is put into it at 0°C?

Picture: The power rating is the work input per unit time. Heat is extracted from the interior of the refrigerator at 0°C since it is extracted from freezing water. To freeze the water, the total latent heat of fusion will have to be extracted from the water. Since the cycle is reversible, the refrigerator is ideal and the quantities of heat exhausted to and extracted from the hot and cold reservoirs are proportional to their absolute temperatures.

Solve:

Write two *algebraic* expressions for the COP, one in terms of Q_c and W and the other in terms of the temperatures of the reservoirs. Remember that $W = Q_h - Q_c$. Also, because the refrigerator is reversible, the heat is proportional to the absolute temperatures of the reservoirs, and the proportionality constant is the same and will cancel out.	
Find an *algebraic* expression for the total heat that must be extracted from the freezing water.	
Find an *algebraic* expression for the work done, in terms of the power and time.	
Substitute the results of the previous two steps into your expression for the COP in terms of Q_c and W. Set the resulting expression equal to the COP expression in terms of the temperatures and solve for time.	
	$t = 902$ s

Check: This refrigerator will take about 15 minutes to freeze the water, which seems a bit fast, but remember that the water starts out at 0°C, so it does not have to be cooled.

Taking It Further: This problem illustrates several common-sense results related to refrigeration. How does the time required to freeze the water vary with each of the following parameters: mass of water, power of the refrigerator, temperature of the outside air?

19.3   The Carnot Engine

In a Nutshell

Carnot's theorem is: No engine working between two given heat reservoirs can be more efficient than a reversible engine working between those reservoirs. In other words, a **reversible** engine is the most efficient.

In order for a process to be **reversible** it must satisfy the following conditions:
1. No mechanical energy can be transformed into thermal energy by friction, viscous forces, or other dissipative forces.
2. Energy transfer as heat can occur only between objects at, or infinitesimally near, the same temperature.
3. The process must be quasi-static so that the system is always in, or infinitesimally near, an equilibrium state.

Most processes we observe in nature are irreversible, so great care must be taken to approximate a reversible system as described above.

A reversible engine working in a cycle between two heat reservoirs is called a **Carnot engine**, and its cycle is called a **Carnot cycle**. The four steps of a Carnot cycle are as follows:
1. **A reversible isothermal expansion.** Prior to this stage the working substance is brought to thermal equilibrium with a high-temperature reservoir of temperature T_h. During this stage the working substance, in thermal contact with a high-temperature reservoir, expands isothermally (at constant temperature). During this expansion it does positive work on an external agent and heat is transferred to it from the high-temperature reservoir.
2. **A reversible adiabatic expansion.** This stage begins when the working substance is thermally isolated from the high-temperature reservoir. During this stage the working substance continues to expand, this time adiabatically; that is, no heat is transferred to or from the material. During this expansion it does positive work on the external agent and the material cools as its internal energy is converted to work. Its temperature continues to drop until it reaches the temperature of the low-temperature reservoir T_c. At this point the working substance is placed in thermal contact with the low-temperature reservoir.
3. **A reversible isothermal compression.** During this stage the working substance, now in thermal contact with the low-temperature reservoir, is isothermally compressed. During the compression the external agent does positive work on it and the working substance rejects heat to the low-temperature reservoir. When the working substance is thermally isolated from the cold-temperature reservoir, this stage ends.
4. **A reversible adiabatic compression.** During this stage the compression continues, this time adiabatically. The external agent does additional positive work on the working substance, which results in its internal energy increasing along with its temperature. The compression continues until its temperature again equals T_h, at which point the engine has completed this stage, the final stage of the cycle. At this moment the working substance is in the exact state it was in when the cycle began at stage 1.

The **efficiency of a reversible engine** (a Carnot engine) can be calculated if an ideal gas is used for the working substance. By Carnot's theorem, this must be the efficiency of every reversible engine and the upper limit possible for any real heat engine operating between the same two temperatures. The efficiency for a reversible heat engine using an ideal gas is calculated to be $\varepsilon_C = 1 - T_c/T_h$, where T_c and T_h are absolute temperatures.

Since the efficiency is the same for all reversible heat engines operating in a cycle, it can be used to define the temperature of the two reservoirs. This allows the definition of a temperature scale that is independent of any particular material or thermometer. To use this definition to construct a temperature scale we choose one fixed point, say the triple point of water, and assign it a value. Then the **thermodynamic temperature** of a reservoir is measured by means of a reversible heat engine operating between the reservoir and a second reservoir maintained at the triple point of water. By measuring the efficiency of this engine we determine the temperature of the reservoir. If the temperature of the triple point of water is given the value of 273.16 K, then the absolute temperature and the ideal-gas temperature will agree over the range for which gas thermometers are able to be used.

If a heat engine has a **second-law efficiency** of 50%, that means the engine has an efficiency equal to 50% of a Carnot engine.

Important Derived Results

Carnot efficiency
$$\varepsilon_C = 1 - \frac{T_c}{T_h}$$

Definition of thermodynamic temperature
$$\frac{T_c}{T_h} = \frac{Q_c}{Q_h}$$

Common Pitfalls

> ➤ All Carnot engines, by definition, have the same efficiency when operating between the same two reservoirs.

5. TRUE or FALSE: In a reversible heat engine, heat must be absorbed or rejected isothermally.

6. In a slow, steady isothermal expansion of an ideal gas against a piston, the work done is equal to the heat input. Is this consistent with the first law of thermodynamics?

Try It Yourself #4

Here is a not very clever idea for a ship's engine: A Carnot engine extracts heat from seawater at 18.0°C and exhausts it to evaporating dry ice, which the ship carries with it, at −78.0°C. If the ship's engines are to run at 8000 horsepower, what is the minimum amount of dry ice it must carry for a single day's running?

Picture: The efficiency of the engine can be determined from the temperatures involved. Determine how much work, in joules, must be done by the engines in one day. The heat rejected to the low-temperature reservoir determines how much dry ice must be sublimated. The latent heat of sublimation of dry ice (carbon dioxide) can be found in Table 18-2 on page 596 of the text.

Solve:

Begin by converting all values to SI units.	
Write an *algebraic* expression for the efficiency of the Carnot engine in terms of the work done, the heat absorbed, and the temperatures of the reservoirs.	
Find an *algebraic* expression for the work, which for this problem is simply power times time.	
Find an *algebraic* expression for the heat the engine absorbs, which is the work it does plus the heat it rejects. The heat rejected to the dry ice is equal to the product of the mass of the ice and the heat of sublimation.	
Finally, substitute expressions into your Carnot efficiency expression. Rearrange *algebraically* to solve for the mass of the dry ice. Substitute values to find the required mass.	$m = 1.8310^6$ kg

Check: This is nearly 2000 metric tonnes of dry ice per day! We said the idea wasn't very clever.

Taking It Further: Can this heat engine ever achieve a 100% efficiency? Why or why not?

Try It Yourself #5

A certain engine has a second-law efficiency of 85.0%. In each cycle it absorbs 480 J of heat from a reservoir at 300°C and rejects 300 J of heat to a cold-temperature reservoir. (a) What is the temperature of the cold reservoir? (b) How much more work could be done by a Carnot engine working between the same two reservoirs and extracting the same 480 J of heat in each cycle?

Picture: The actual efficiency can be calculated using the information provided. The Carnot efficiency can then be calculated and used to determine the temperature of the low-temperature reservoir.

Solve:

Write an *algebraic* expression for the actual efficiency of the engine in terms of the heats extracted from and transferred to the reservoirs.	
Write an expression for the Carnot efficiency.	
We know the actual efficiency is 85% of the Carnot efficiency. Use this to solve for the temperature of the cold reservoir.	$T_{\rm c} = 47°C$
Now find the work done by an ideal Carnot engine operating between the same temperatures, using the expression for the Carnot efficiency.	
The work done by the engine in this problem is 85% of the work done by a Carnot engine, so we can calculate the extra work that would be done by the Carnot engine.	$W_{\rm additional} = 32.0$ J

Check: The Carnot engine better do more work, as it is by definition the most efficient engine.

Taking It Further: What would make an engine operate at less efficiency than a Carnot engine?

19.4   Heat Pumps*

In a Nutshell

A **heat pump**, which is essentially a refrigerator, is used to pump heat from a colder region (for example, outdoors) to a warmer region (for example, the interior of a building). For a heat pump we are interested in the heat Q_h delivered into the high-temperature reservoir, the house. Consequently the parameter of interest is usually not the COP (Q_c/W) but the ratio Q_h/W, where the work W is the amount of energy that will appear on our electric bill. The parameter of interest for a heat pump can be shown to equal 1 plus the coefficient of performance.

Important Derived Results

Coefficient of performance for heat pump $\text{COP}_{\text{HP}} = \dfrac{Q_h}{W} = 1 + \text{COP}$

Maximum coefficient of performance for $\text{COP}_{\text{HP max}} = \dfrac{T_h}{T_h - T_c} = \dfrac{T_h}{\Delta T}$
heat pump

19.5   Irreversibility, Disorder, and Entropy

In a Nutshell

Entropy S is a thermodynamic quantity that can be thought of as a measure of the disorder of a system. It is defined as $dS = dQ_{\text{rev}}/T$, where dQ_{rev} is the energy that must be transferred to the system as heat in a reversible process. Reversible heat transfer is not required, however, for a change in entropy to occur. Changes in entropy, like changes in internal energy, depend only on the initial and final states of the system, and not on the process taking the system from the initial state to the final state. Thus entropy, like internal energy, is a state function.

During a reversible process, the entropy change of the universe is zero.
During an irreversible process, the entropy of the universe increases.
For any process, the entropy of the universe never decreases.

The entropy change for an ideal gas can be shown to be $\Delta S = \int (dQ_{\text{rev}}/T) = C_v \ln(T_2/T_1) + nR \ln(V_2/V_1)$, where T_1, T_2, V_1, and V_2 are the initial and final temperatures and volumes of the gas, respectively.

Physical Quantities and Their Units

Entropy   units of joules/kelvin

---

*Optional material.

Important Derived Results

Entropy

$$dS = \frac{dQ_{\text{rev}}}{T}$$

Entropy change for an ideal gas

$$\Delta S = \int (dQ_{\text{rev}}/T) = C_{\text{v}} \ln(T_2/T_1) + nR \ln(V_2/V_1)$$

Common Pitfalls

> ➤ Changes in entropy may be calculated only for reversible processes. Entropy is a function of state and a given entropy change is associated with a given change in state, but not with any specific process bringing the system from the initial state to the final state. To calculate the entropy change, we must calculate the entropy change for any reversible process bringing the system from the initial state to the final state.

> ➤ Note that dQ_{rev}/T does not equal the entropy of the initial or final state in a process but only the *difference* in the entropy between the two states. It is rather like potential energy in that only differences, or changes, are defined.

> ➤ The second law does not say that the entropy of some object or system cannot decrease; in fact, it can decrease. But the decrease will always be made up (or more than made up) by an increase elsewhere. The second law states that the total entropy of the *universe* may not decrease.

> ➤ In calculating the change in the entropy of a system, you must always use the absolute temperature scale.

> ➤ In calculating the change in the entropy of a system, you must carefully consider whether the heat is added to or removed from the various parts of the system. If heat is removed, then the entropy is reduced. If heat is added, then the entropy is increased.

7. TRUE or FALSE: In any adiabatic process, the entropy change of the system is zero.

8. If a gas expands freely into a larger volume in an insulated container so that no heat is added to the gas, its entropy increases. In view of the definition of ΔS, how can this be?

Try It Yourself #6

When 1 kg of water is frozen under standard conditions, by how much does its entropy change?

Picture: Change in entropy is equal to heat absorbed divided by the temperature. "Standard conditions" implies a temperature of 0°C.

Solve:

Write an *algebraic* expression for the change in entropy of the water.	

Find the heat extracted from the water. You will need the latent heat of fusion for water.	
Solve for the change in entropy.	$\Delta S = -1221$ J/K

Check: The change in entropy is negative, indicating that the ordering of the water has increased. Frozen water is, in fact, more ordered than liquid water, which is much more free to flow, so this makes sense.

Taking It Further: If the entropy of the universe cannot decrease, how is the decrease in entropy of the water possible?

19.6   Entropy and the Availability of Energy

In a Nutshell

During an irreversible process, energy equal to $T\,\Delta S_{\mathrm{u}}$ becomes unavailable to do work, where T is the temperature of the coldest available reservoir, and ΔS_{u} is the change in the entropy of the universe. We call this energy that becomes unavailable to do work the "lost work."

Important Derived Results

"Lost work" $W_{\mathrm{lost}} = T\,\Delta S_{\mathrm{u}}$

Common Pitfalls

9. TRUE or FALSE: When an irreversible process takes place, the universe becomes more disordered.

10. Heat flows from a hotter to a colder object. By how much is the entropy of the universe changed? In what sense does this correspond to energy becoming unavailable for doing work?

Try It Yourself #7

In a vacuum bottle, 350 g of water and 150 g of ice are initially in equilibrium at 0°C. The bottle is not a perfect insulator. Over time, its contents come to thermal equilibrium with the outside air at 25°C. How much does the entropy of universe increase in this process?

Picture: The total change in entropy of the universe will equal the increase in entropy of the ice and water as they warm, plus the decrease in entropy of the surrounding air as it loses heat.

Solve:

First find an *algebraic* expression for the change in entropy of the ice as it melts. This happens at a constant temperature, and involves the latent heat of fusion.	
Now find an *algebraic* expression for the change in entropy of the water as it is heated. Because the temperature is *not* constant during this process, the integral is required. The differential amount of heat absorbed as the ice increases its temperature a small amount dT involves the specific heat of water.	
Find the change in entropy of the air, which is equal to the heat lost in melting the ice plus the heat lost in warming the water, divided by the constant temperature of the air.	
The change in entropy of the universe is the sum of these changes in entropy.	$\Delta S_{\mathrm{u}} = 23.1$ J/K

Check: The entropy of the universe increases, which it must.

Taking It Further: Does this change in entropy depend on how the ice is melted?

Try It Yourself #8

The interior of a refrigerator's freezing compartment is at 10°F. The kitchen is at 78°F. Suppose that heat leaks through the walls into the freezing compartment at a rate of 70.0 cal/min. (a) In 1 hour, how much has the entropy of the universe been increased by this heat leakage? (b) How much energy becomes unavailable for doing work when this heat leaks into the freezer compartment?

Picture: The total change in entropy of the universe will equal the increase in entropy of the freezer as it warms, plus the decrease in entropy of the surrounding air as it loses heat. From the change in entropy you can find the "lost work."

Solve:

Convert the temperatures to the Kelvin scale and the rate of heat leakage to J/s.	
Find the entropy change of the freezing compartment which increases by an amount equal to the heat delivered to it divided by its absolute temperature.	
In the same way, obtain an expression for the entropy decrease of the kitchen outside the refrigerator.	
The entropy change of the universe is that of the freezing compartment plus that of the outside world.	$\Delta S_u = 8.52$ J/K
Relate the "lost work" to the entropy change and the temperature of the low-temperature reservoir.	$W_{lost} = 2.22$ kJ

Check: The change in entropy of the universe and the "lost work" are both positive, which they must be.

Taking It Further: How can the entropy of the surrounding air decrease?

19.7 Entropy and Probability

In a Nutshell

The second law of thermodynamics is not absolute—rather, it is a statement of probability. When applied to small numbers of molecules, there is a reasonable probability of finding the molecules in a relatively low entropy state. But that probability is smaller than the probability of finding the molecules in a relatively high entropy state. However, *thermodynamics itself is applicable only to macroscopic systems with large numbers of molecules*. As a consequence, the probability that any system is found in a relatively lower entropy state is vanishingly small.

QUIZ

1. TRUE or FALSE: All heat engines operating between the same two heat reservoirs have the same efficiency.

2. TRUE or FALSE: The transfer of heat across a finite temperature difference is an irreversible process.

3. Why do engineers designing a steam-electric generating plant always try to design for as high a feed-steam temperature as possible?

4. The conduction of heat across a temperature difference is an irreversible process, but the object that lost heat can always be rewarmed, and the one that gained it can be recooled. The dissipation of mechanical energy, as in the case of an object sliding across a rough table and slowing down, is irreversible, but the object can be cooled and set moving again at its original speed. So in just what sense are these processes "irreversible"?

5. In discussing the Carnot cycle, we say that extracting heat from a reservoir isothermally does not change the entropy of the universe. In a real process, this is a limiting situation that can never quite be reached. Why not? What is the effect on the entropy of the universe?

6. Consider an engine in which the working substance is 1.23 mol of an ideal gas for which $\gamma = 1.41$. The engine runs reversibly in the cycle shown on the PV diagram. The cycle consists of an isobaric (constant pressure) expansion a at a pressure of 15.0 atm, during which the temperature of the gas increases from 300 K to 600 K, followed by an isothermal expansion b until its pressure becomes 3.00 atm. Next is an isobaric compression c at a pressure of 3.00 atm, during which the temperature decreases from 600 K to 300 K, followed by an isothermal compression d until its pressure returns to 15 atm. Find the work done by the gas, the heat absorbed by the gas, the internal energy change, and the entropy change of the gas, first for each part of the cycle and then for the complete cycle. You may need to show your work on a separate sheet of paper.

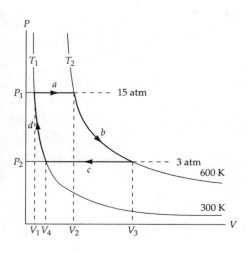

7. A heat engine works in a cycle between reservoirs at 273 K and 490 K. In each cycle the engine absorbs 1250 J of heat from the high-temperature reservoir and does 475 J of work.
   (a) What is its efficiency? (b) What is the change in entropy of the universe when the engine goes through one complete cycle? (c) How much energy becomes unavailable for doing work when this engine goes through one complete cycle?

Chapter 20

Thermal Properties and Processes

20.1   Thermal Expansion

In a Nutshell

When the temperature of a solid or liquid increases, it usually expands. When a bar of length L is heated, raising its temperature by ΔT, the length of the bar changes by ΔL. The fractional change in the length is $\Delta L/L$. The ratio of the fractional change in the length of an object to the change in temperature is called the **coefficient of linear expansion** α.

Similarly, the **coefficient of volume expansion** β is defined as the ratio of the fractional change in volume to the change in temperature (at constant pressure). For a given material, the relation between the coefficient of volume expansion and the coefficient of linear expansion is $\beta = 3\alpha$.

Physical Quantities and Their Units

Coefficient of expansion, α or β units of 1/K

Important Derived Results

Coefficient of linear expansion $\alpha = \lim\limits_{\Delta T \to 0} \dfrac{\Delta L/L}{\Delta T} = \dfrac{1}{L}\dfrac{dL}{dT}$

Coefficient of volume expansion $\beta = \lim\limits_{\Delta T \to 0} \dfrac{\Delta V/V}{\Delta T} = \dfrac{1}{V}\dfrac{dV}{dT}$

Common Pitfalls

> ➤ In problems involving thermal expansion, the units of temperature must be consistent with those of α or β. On the other hand, any units may be used for L and ΔL (or V and ΔV) as long as they are the same. As always, include units with your numerical quantities and check that the units on both sides of an expression match before doing the calculation.

1. TRUE or FALSE: All materials expand when heated.

2. A metal plate with a circular hole drilled through it is uniformly heated. As the plate gets hotter, it expands. Does the hole get bigger or smaller? Why?

Try It Yourself #1

A brass pin is exactly 5.00 cm long when it is at a temperature of 140.0°C. What is its length when it cools to 20.0°C?

Picture: Look up the coefficient of thermal expansion for brass in Table 20-1 on page 666 of the text and use it to determine the new length.

Solve:

The coefficient of thermal expansion is the fractional change in length per kelvin. Use this to find an *algebraic* expression for the change in length.	
Find the final length, which will be equal to the initial length plus the change in length.	$L_{\text{final}} = 0.0499$ m

Check: Because the pin cools, the final length should be smaller than the initial length.

Taking It Further: In practice, will you notice this change in length? Why or why not?

Try It Yourself #2

A 10.0-L Pyrex flask is filled to the brim with acetone at 12.4°C. To what temperature must the flask and its contents be heated, or cooled, so that 85.0 cm^3 of acetone overflow the flask?

Picture: Obtain an expression for the change in the volume of the flask and the acetone in terms of the temperature change. You will need the coefficients of thermal expansion for Pyrex glass and acetone found in Table 20-1 on page 666 of the text. Remember the volume coefficient of thermal expansion is three times the linear coefficient of thermal expansion. You want the difference in the change of these volumes to be 85.0 cm^3.

Solve:

Obtain a generic *algebraic* expression for the change in volume due to the temperature change.	

Apply the expression above to both the change in volume of the flask and the acetone. Set the difference of the change in volume of the flask and the acetone equal to 85.0 cm^3 and solve for the temperature change.	
From the initial temperature and the temperature change, find the final temperature required.	$T = 18.1°C$

Check: A temperature change of almost 6 K seems reasonable.

Taking It Further: If instead of acetone, the Pyrex flask were filled with a solid piece of brass, what would happen when the flask was heated? What if the flask were filled with a solid piece of Invar?

20.2   The van der Waals Equation and Liquid–Vapor Isotherms

In a Nutshell

The van der Waals equation of state describes the behavior of gases over a wide range of pressures more accurately than does the ideal-gas equation of state $PV = nRT$. The van der Waals equation of state for n moles of a gas is $\left(P + an^2/V^2\right)(V - bn) = nRT$. The constants a and b are determined experimentally and provide for the attractive forces between molecules and the finite size of the molecules, respectively.

The pressure at which a liquid coexists in equilibrium with its own vapor is called the **vapor pressure**. The temperature at which the vapor pressure equals 1 atm is called the **normal boiling point**.

At temperatures greater than the **critical temperature**, a gas will not condense at any pressure.

Important Derived Results

The van der Waals equation of state
$$\left(P + \frac{an^2}{V^2}\right)(V - bn) = nRT$$

Common Pitfalls

> At high pressures or low temperatures, the ideal-gas law breaks down, and the van der Waals equation should be used instead.

3. TRUE or FALSE: The constant b that appears in the van der Waals equation is related to the volume of the molecules themselves.

4. Under what conditions does the van der Waals equation of state reduce to the ideal-gas law?

Try It Yourself #3

Fifty moles of steam are confined in a 30.0-L container at 600°C. (a) Calculate the pressure using the ideal-gas equation. (b) The van der Waals constants for steam are $a = 5.43$ atm $\cdot$ L^2/mol^2 and $b = 0.03$ L/mol. Calculate the pressure using the van der Waals equation. Actual measurements show the pressure to be 113 atm.

Solve:

Calculate the pressure using the ideal-gas equation.	
	$P_{\text{ideal}} = 119$ atm
Calculate the pressure using the van der Waals equation.	
	$P_{\text{vdW}} = 111$ atm

Check: The van der Waals pressure agrees more closely with the experimentally measured pressure than the ideal-gas pressure. Both, however, are within 5% of the experimental value, demonstrating that quite high pressures are needed for the difference in these two expressions to be significant.

20.3 Phase Diagrams

In a Nutshell

The equilibrium state of a substance held at constant volume depends on its temperature and pressure. A plot of pressure versus temperature for such a substance is called a **phase diagram**, where the state of a substance can be represented by a point. On a phase diagram, one region represents states of solid phase, another region represents states of liquid phase, and yet another region represents states of gaseous phase.

The point on a phase diagram where the liquid and vapor phases have the same density is called the **critical point**, and the temperature at this point is called the **critical-point temperature** T_c. At this point and above it, there is no distinction between the liquid and gas phases. Every substance has a unique **triple point**, the point at which the vapor, liquid, and solid phases can coexist in equilibrium. The triple-point temperature for water is 273.16 K = 0.01°C and the triple-point pressure is 4.58 mmHg.

20.4 The Transfer of Heat

In a Nutshell

Temperature differences result in the transfer of energy from one place to another by three processes: conduction, convection, and radiation. In **conduction**, heat is transferred by interactions among atoms or molecules, though there is no transport of the atoms or molecules themselves. In **convection**, heat is transferred via mass transport. This occurs, for example, in a flowing fluid. In **radiation**, heat is transported via electromagnetic radiation that is emitted and absorbed. In all three processes the net flow of heat (that is, the transfer of energy) is from a region of higher temperature to a region of lower temperature.

When heat flows by conduction along a solid bar, the rate at which it flows is called the **thermal current** I, measured in units such as joules per second. The thermal current is related to the **temperature gradient** $\Delta T / \Delta x$ by $I = |\Delta Q / \Delta t| = kA |\Delta T / \Delta x|$, where ΔQ is the net flow of heat during time Δt. A is the cross-sectional area of the bar, and the constant k is the **coefficient of thermal conductivity** of the substance.

If the **thermal resistance** of a material is defined as $R = |\Delta x| / (kA)$, where Δx is the thickness of the material, then we can write $\Delta T = IR$. This expression is analogous to the expression for the viscous flow of a fluid through a pipe. In Chapter 25 we will find a similar expression involving electrical current and resistance.

If heat flows from a warmer region to a cooler region through a series of slabs, the equivalent thermal resistance R_{eq} of the series is the sum of the thermal resistances of the individual slabs: $R_{eq} = R_1 + R_2 + \ldots$.

Similarly, if heat is conducted from a warmer region to a cooler region through two or more . parallel paths, the reciprocal of the equivalent thermal resistance of the paths is the sum of the reciprocals of the thermal resistances of the individual paths—that is, $1/R_{eq} = 1/R_1 + 1/R_2 + \ldots$

The heat transferred to or from an object by convection is approximately proportional to the surface area of the object and to the difference in temperature between the object and the surrounding fluid.

The rate at which a surface radiates heat is given by the **Stefan-Boltzmann law**: $P_r = e\sigma AT^4$, where e is the **emissivity** of the surface ($0 \leq e \leq 1$), A is the area of the surface, T is the surface temperature in kelvins, and σ is Stefan's constant; $\sigma = 5.6703 \times 10^{-8}$ W/(m$^2 \cdot$ K^4).

The rate at which a body absorbs radiant heat from an object is $P_a = e\sigma AT_0^4$, where T_0 is the temperature of the radiating object. Therefore the net power radiated by an object at temperature T in an environment at temperature T_0 is given by $P_{net} = e\sigma A(T^4 - T_0^4)$. Note that a good emitter (an object having a high emissivity e) is also a good absorber.

An object that absorbs all the radiation incident upon it has an emissivity equal to 1 and is called a **blackbody**. A blackbody is also an ideal radiator. When a body emits thermal radiant energy, it does so over a continuum of wavelengths. The wavelength λ_{max} at which a blackbody emits radiant energy at the greatest rate is inversely proportional to the absolute temperature of the body. This result is known as Wien's displacement law: $\lambda_{max} = (2.898 \text{ mm} \cdot \text{K})/T$.

All three mechanisms of heat flow are driven by a difference in temperature. Independent of which of the mechanisms are at work, the rate of cooling of a warm body is approximately proportional to the temperature difference between the body and its surroundings. This result, called **Newton's law of cooling**, holds whether heat is being transferred by conduction, convection, radiation, or some combination of the three. It is established by applying the differential approximation to the various heat transfer mechanisms and is most accurate when the temperature differences are small.

Physical Quantities and Their Units

Stefan's constant $\qquad\qquad\qquad\qquad\qquad$ $\sigma = 5.6703 \times 10^{-8}\ \mathrm{W/(m^2 \cdot K^4)}$

Important Derived Results

Thermal conduction $\qquad\qquad\qquad\qquad$ $I = \left| \dfrac{\Delta Q}{\Delta t} \right| = kA \left| \dfrac{\Delta T}{\Delta x} \right| = \dfrac{|\Delta T|}{R}$

Thermal resistance $\qquad\qquad\qquad\qquad$ $R = \dfrac{|\Delta x|}{kA}$

Resistance in series $\qquad\qquad\qquad\qquad$ $R_{\mathrm{eq}} = R_1 + R_2 + \ldots$

Resistance in parallel $\qquad\qquad\qquad\qquad$ $\dfrac{1}{R_{\mathrm{eq}}} = \dfrac{1}{R_1} + \dfrac{1}{R_2} + \ldots$

Stefan-Boltzmann law $\qquad\qquad\qquad\qquad$ $P_{\mathrm{r}} = e\sigma A T^4$

Absorption of radiation $\qquad\qquad\qquad\qquad$ $P_{\mathrm{a}} = e\sigma A T_0^4$

Net radiative thermal current $\qquad\qquad\qquad$ $P_{\mathrm{net}} = e\sigma A \left(T^4 - T_0^4 \right)$

Wein's displacement law $\qquad\qquad\qquad\qquad$ $\lambda_{\mathrm{max}} = \dfrac{2.898\ \mathrm{mm} \cdot \mathrm{K}}{T}$

Common Pitfalls

> ➤ There are three and only three mechanisms for the transfer of energy via heat: conduction, convection, and radiation. Remember that, by definition, heat is the transfer of energy due to a difference in temperature.

> ➤ In conduction and convection problems, only temperature differences matter, so Celsius degrees and kelvins may be used interchangeably. In radiation problems this is not true; you must use absolute temperatures. Remember, heat transfers are always from warmer regions to cooler regions.

5. TRUE or FALSE: Energy can be transferred even without any mass physically moving from one place to another.

6. Materials used commercially for building insulation tend to have relatively little mass and occupy a lot of space: they are foamy, porous materials or masses of compacted fibers or som such. Why?

Try It Yourself #4

The figure shows a system holding liquid helium at 4.20 K (the boiling point of helium). A cylindrical can 5.00 cm in diameter and 7.00 cm high is held away from the walls, which are at 77.0 K (the boiling point of nitrogen) by two stainless steel pins . The space between the can and the walls is evacuated. The steel pins are each 1.00 mm in diameter and 6.00 cm long and have a thermal conductivity of 13.4 W/(m · K). If the latent heat of vaporization of helium is 21.0 kJ/kg, at what rate does the helium boil off? (Assume that the emissivity of the can's exterior is 0.250.)

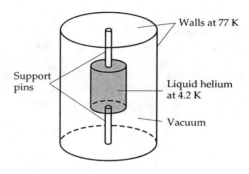

Picture: The heat that evaporates the helium is delivered to it through the support pins and by radiation from the 77-K walls. Thus the rate at which the helium must absorb heat is equal to the conductive thermal currents through the pins plus the radiative thermal current.

Solve:

Find an *algebraic* expression for the rate at which the helium absorbs heat to evaporate. This will be in terms of the mass and latent heat of vaporization.	
Find an *algebraic* expression for the rate at which thermal energy is transferred via conduction through the pins.	
Find an *algebraic* expression for the rate at which thermal energy is transferred radiatively to the helium.	
Equate the rate at which the helium absorbs energy to the sum of the previous two steps. Solve for the rate at which the helium boils off, dm/dt.	$\dfrac{dm}{dt} = 1.57 \times 10^{-6}$ kg/s

Check: This seems to be a pretty slow rate of evaporation, which is a good thing since liquid helium is pretty expensive.

Taking It Further: How could you most effectively slow the boiling off of the helium?

Try It Yourself #5

In a certain experiment, heat is transferred from a source at 227°C to water at 30.0°C through a rod 3.00 cm in diameter, as shown. The rod consists of a 12.0-cm-long aluminum rod butted end to end with a 5.00-cm-long copper rod. What is the equilibrium temperature of the aluminum-copper junction if the aluminum rod is in thermal contact with the 227°C heat source and the copper rod is in contact with the 30.0°C water bath? (The thermal conductivities can be found in Table 20-4 on page 676 of the text.)

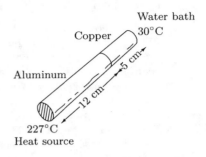

Picture: The thermal current is related to the thermal conductivities, cross-sectional areas, and the temperature gradients of each rod. Equilibrium will be reached when the thermal currents from each rod are equal.

Solve:

Find an *algebraic* expression for the thermal current in each rod.	
Equate the two expressions for the thermal current and solve for the temperature at the junction.	
	$T = 68.9°C$

Check: This seems reasonable. The temperature should be between the temperature of the two baths.

Taking It Further: If the rod is reversed so that the copper is in contact with the heat source and the aluminum is in contact with the water bath, will the temperature of the junction rise or fall? Why?

QUIZ

1. TRUE or FALSE: The constant a that appears in the van der Waals equation is related to the volume of the molecules themselves.

2. The ordinary thermometers we see every day are mostly alcohol in glass. Would the use of such a thermometer be practical if alcohol and glass had the same thermal coefficient of volume expansion?

3. If an ordinary (liquid-in-glass) thermometer is placed in something quite hot, the liquid column may actually drop a little before it starts to rise. What is going on here?

4. When you are trying to open a glass jar of food with a stuck lid, it often helps to run hot water over the metal lid for a little while. Why?

5. Vessels like vacuum bottles or dewar flasks, which are designed to keep fluids very cold, are made with double-glass walls. The inner surfaces are silvered and there is a vacuum between the walls. Discuss how this design minimizes heat losses.

6. A vacuum bottle with an inside diameter of 6.00 cm contains 150.0 g of water and 75.0 g of ice at 0.00°C. Heat leakage through the walls of the bottle is negligible, but it is closed with a round cork stopper 2.00 cm thick whose thermal conductivity is 0.500 W/(m · K). If the surroundings are at 28.0°C, how long does it take for all the ice to melt?

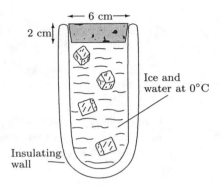

7. My car has a 40.0-L gasoline tank. If I fill it completely full of gasoline at a temperature of 12°C and then let the car sit in the Sun until the temperatures of the gasoline and the tank reach 30°C, how much gasoline spills out of the tank if I forget to close the cap? The thermal coefficient of volume expansion for gasoline is 9×10^{-4} K^{-1} and the tank is made of steel.

Answers to Problems

Chapter 1

COMMON PITFALLS

1. True. Because the meter is defined as the distance light travels in some small fraction of a second, if the duration of the second doubled, the length of the meter would also double.

2. Certainly. It could have fundamental units for time and speed. The unit for length could then be defined in terms of these fundamental units.

3. False. In order not to change the value of the quantity the conversion factor must have a size of 1; however, the conversion factor must have units–otherwise it is useless.

4. The SI unit for speed is the meter per second. We determine the conversion factor as follows:

$$\left(\frac{1 \text{ furlong}}{\text{fortnight}}\right)\left(\frac{1 \text{ fortnight}}{14 \text{ days}}\right)\left(\frac{1 \text{ day}}{24 \text{ h}}\right)\left(\frac{1 \text{ h}}{3600 \text{ s}}\right)\left(\frac{1 \text{ mi}}{8 \text{ furlongs}}\right)\left(\frac{1610 \text{ m}}{1 \text{ mi}}\right) = 1.66 \times 10^{-4} \text{ m/s}$$

The conversion factor is:

$$\frac{1.66 \times 10^{-4} \text{ m/s}}{1 \text{ furlong/fortnight}}$$

5. True. In order not to change the value of the quantity the conversion factor must have a size of 1. It must also be dimensionless. If, for example, a conversion factor had dimensions of L/T, then multiplying a time by this conversion factor could change a time into a distance! Clearly it doesn't make sense for time and distance to be the same quantity, so all conversion factors are dimensionless.

6. To be added, the quantities have to have the same dimensions but not necessarily the same units. For example, if you add 3 meters and 72 centimeters, you get 3 meters 72 centimeters. Normally, you will want to convert the units of one of the quantities to the units of the other before adding them. Quantities can be divided even when they have different dimensions, as when distance is divided by time to get speed.

7. False. Any finite-sized number raised to the zero power is 1.

8. You are dividing a number with four significant figures by a number with three significant figures. The result should therefore have three significant figures. It should be written 55.0.

9. False. The components of a vector are scalar numbers. The components can be turned into vectors by multiplying them by the appropriate unit vector $\hat{\imath}$, $\hat{\jmath}$, or $\hat{k}$.

10. In general, both the magnitude of a vector and its components have dimensions associated with them. They are the same dimensions as the vector itself. A vector can be written as its magnitude multiplied by a unit vector. Since the unit vector is dimensionless, the magnitude must have the required dimensions.

TRY IT YOURSELF–TAKING IT FURTHER

3. There are several possibilities. Here are a few: km/h, cm/min, ft/s, light years/h, furlongs/fortnight.

4. If you multiply ft/ns by a time in ns, units analysis shows that you will be left with units of ft, which is a distance. So you should multiply the speed by the time, with the appropriate units.

7. Assume you take 1.5 steps every second. Then it would take 1 000 000 seconds, or approximately 11–12 days.

8. It would be in the second quadrant. Both vectors would have negative $\hat{\imath}$ components, but the positive $\hat{\jmath}$ component of $\vec{A}$ is larger than the negative $\hat{\jmath}$ component of $-\vec{B}$.

QUIZ

1. False

2. True

3. They must be dimensionless so as to not change the kind of measurement being made. For instance, you do not want to change a pressure into a velocity. However, units are required to make the conversion of pressure from units of atmospheres to torr.

4. v^2 has dimensions of L^2/T^2. As a result $v^2 = 2ad$ is the dimensionally correct form.

5. $\vec{C} = -18.6\hat{\imath} + 74.5\hat{\jmath}$, $C = 76.8$, $\hat{C} = -0.242\hat{\imath} + 0.970\hat{\jmath}$, $\theta = 104°$

Chapter 2

COMMON PITFALLS

1. False. It is possible that the distance traveled is 60 km. However, one can also envision a situation where the velocity is greater than +60 km/h for enough time that the car goes past the 60 km point. Then the car reverses directions so that at one hour the car's *displacement* is 60 km. However, it actually traveled a greater distance because it had to backtrack.

2. No. Average speed is defined as the total distance traveled divided by the time interval required for that travel. Since both quantities are by definition positive, the average speed must also be positive.

3. True. They are both first derivative relations.

4. No. If a particle is already moving in the negative direction, a negative acceleration means the particle is speeding up in the negative direction.

5. False. Neglecting air resistance, an object in free-fall near the surface of the earth always has an acceleration of 9.8 m/s^2 directed toward the center of the earth.

6. In both situations the bolt will experience constant-acceleration motion, with the same acceleration in the downward direction. When the elevator is moving up, the bolt has an initial velocity upward, so it will continue traveling upward, but accelerate downward until it slows to a momentary stop when it continues to accelerate downward and begins the downward portion of its path. When the elevator is moving down, the bolt will simply accelerate downward, continually gaining speed. Both bolts will hit the bottom of the shaft with the same final velocity because they both have the same acceleration, the same displacement, and the same magnitude of initial velocity.

7. True. Differentiating will show that the velocity is A.

8. The area "under" a velocity vs. time graph between the initial and final times is the displacement. If the velocity is negative over some time interval, then the area, and hence displacement, is negative in that region. The instantaneous acceleration is given by the time derivative of the velocity, or the slope of the curve.

TRY IT YOURSELF–TAKING IT FURTHER

1. In this problem the displacement is all in the same direction, so the displacement and the distance traveled are identical.

2. In this problem the distance traveled (≈ 540 m) is much greater than the displacement (≈ 370 m). The average speed is total distance traveled divided by the total time required. ave speed = 0.18 m/s.

3. No, but the acceleration is *always* the second derivative of the position with respect to time. If the position varies linearly with time, for instance, the acceleration is zero.

4. Use the expressions from step three in the solution, but set the final velocity equal to zero. The displacement you find will have to be added to the first 300 m the rocket traveled to find the maximum height, which is 1200 m.

5. Al will have to have a very large initial acceleration, reach some maximum speed before meeting Bob, and then slow continuously as he catches up to Bob. Whether or not this is desirable or possible depends on the maximum acceleration Al can achieve, and whether or not he can drive safely at the maximum speed.

6. When we take the derivative of the velocity with respect to time, the resulting expression still has a t in it, so it cannot be constant.

7. You need to integrate twice. The first time, find the *indefinite* integral of the acceleration, which gives you velocity. The integration constant is the initial velocity. Then integrate the resulting velocity expression from the initial time to the final time to find the total displacement.

QUIZ

1. False. The average velocity depends only on the initial and final positions and not on the distance traveled.

2. False. If a particle is moving in the negative direction and is slowing down, its velocity is negative while its acceleration is positive.

3. If the acceleration is constant, $v_{av} = \frac{1}{2}(v_1 + v_2) = 13$ m/s and $\Delta x = v_{av}t = 13$ m. If you do not know whether or not the acceleration is constant, you cannot determine these quantities.

4. The bolt continues to ascend until its velocity reaches zero.

5. It takes a sprinter several seconds to reach top speed. Thus in the first three seconds he is traveling slower than 10 m/s most of the time. He can't average 10 m/s over 40 s because a sprinter is unable to sustain top speed for that length of time.

6. (a) 15.2 mi, (b) 3.03 mi/h, (c) 5.37 mi/h

7. 14.5 m/s

Chapter 3

COMMON PITFALLS

1. True. This is the definition of displacement.
2. The two vectors must be antiparallel and have the same magnitude.
3. True. It always points in the direction of $\vec{r}_2 - \vec{r}_1$, two successive, infinitesimally close position vectors.
4. When you "sight in" a rifle at a particular distance, you are adjusting the sights to compensate for the distance the bullet falls below its initial line of flight. The axis of the barrel thus points above the sight line. When the target is closer than the distance for which the rifle was sighted, the bullet will fall a shorter distance. If you aim at the target, the bullet will be high when it gets there, so you should aim *below* the target to compensate.
5. False. The instantaneous-acceleration vector points radially inward toward the center of the circular path being followed. Its magnitude is equal to v^2/r.
6. False. The average-acceleration vector for a time interval is in the same direction as the change in the *velocity* vector for that time interval—radially inward. The position vector changes along the circumference of the circle being traversed.

TRY IT YOURSELF–TAKING IT FURTHER

1. Your average velocity will be the same because the net displacement is still the same, as is the time required to undergo that displacement. Your average speed, however, will decrease.
2. Maximum speed with respect to the earth would have the swimmer pointed directly downstream. Clearly this won't get her across the river. If she points almost completely downstream, with just a small component of her velocity directed across the river, she will eventually cross. But it will take her a long time, and she will have traveled a considerable distance downstream.
3. As the table is tipped downward the ball can still hit the same point. However, to do so the velocity of the ball must increase dramatically until at approximately 19° the angle is so steep that no initial velocity will allow the ball to hit the original location.
4. The receiver can run more slowly. If the ball is thrown at an angle greater than 50°, the horizontal component of the ball's velocity is smaller, so it travels a smaller horizontal distance. In addition, it takes the ball longer to travel this now shorter horizontal distance. If the ball is thrown at an angle between 40° and 50°, the range will increase, but more slowly than the required time for flight, so the receiver can still run more slowly.
5. Yes. If you apply your brakes lightly at first, and more heavily as you progress down the off ramp, then you will slow down a small amount initially, and you will have a greater tangential acceleration as you come to a stop, when you are traveling more slowly, and your centripetal acceleration is smaller.
6. Take-off and landing could generate a few g's of acceleration, but sharp turns and "loop-the-loop" maneuvers generate the highest accelerations, and so are the most dangerous.

QUIZ

1. True
2. True
3. (a) Straight-line motion toward or away from the origin. (b) The particle is moving in a straight line with variable speed. (c) The particle is traveling in a circle. (d) However the position of the particle changes, the particle is moving with a constant speed.
4. The direction of motion (the velocity vector) is changing, so there must be an acceleration causing the change in the velocity vector.
5. Your motion diagram should resemble the one shown here. The acceleration is always 9.81 m/s^2 "downward". The horizontal component of the velocity is always the same. The vertical component is reduced on the way up, and increases on the way back down.

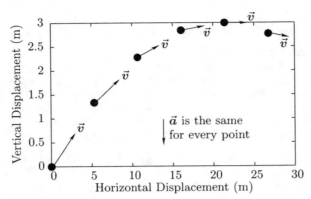

6. 9.12 m/s
7. 31.2 ft/s

Chapter 4

COMMON PITFALLS

1. False. Several different forces may be acting on the object. It is the vector sum of all the forces acting on it that must be zero.
2. You are not actually thrown outward. You and your seat are initially moving forward at the same speed. When the car makes a sharp turn—say to the left—your seat moves with it, and you are pulled to the left literally by the seat of your pants. The inertia of your head and upper torso cause them to lag behind, and it seems as if you are being thrown to the right.
3. True.
4. Microscopic extensions of your hand and the carpet become intertwined. In order for your hand to move over the carpet, the outer electrons in the molecules of your hand repel the outer electrons of the molecules of the carpet to "push" them out of the way. But the carpet also pushes back. This is the fundamental source of friction.
5. True. Just apply Newton's second law to find the mass from the measured force and acceleration.
6. No. Knowing the direction of the force will give you only the direction of the acceleration, *not* the direction of the velocity.
7. False. The astronaut's *apparent* weight is zero. The true weight, which is the gravitational force exerted on the astronaut by Earth, is the force that keeps the astronaut moving in a curved orbit around Earth.
8. No. Your weight, which is the force of Earth's gravity on you, remains essentially constant. Your *apparent* weight—the force of the floor of the elevator on you, does increase as the elevator accelerates upward, and decrease as the elevator accelerates downward.
9. False. As long as the spring is not stretched past its elastic limit, the force is proportional to the distance stretched or compressed. So when the spring is stretched twice the distance, the spring will exert twice the force, not half.
10. This statement is false. When a spring is compressed, it always exerts a force in the direction such that it will return to its equilibrium position. Whether or not this force is in a positive or negative direction depends on the coordinate system you have chosen.
11. False. Knowledge of all the forces and the mass specifies only acceleration. However, knowledge of the initial position and velocity is also needed to completely specify the motion.
12. The forces acting on the car are the gravitational force of Earth, the contact force of the road, and the contact force of the air (air resistance). Because the magnitude and direction of the velocity are constant, the acceleration must be zero. Because the acceleration is zero, it follows from Newton's second law ($\vec{F} = m\vec{a}$) that the net force is also zero.
13. False. These forces both act on the same object, so they can't possibly be an action–reaction pair. Together they constitute the net force acting on the object. Because the acceleration is zero, it follows from Newton's second law, $\vec{a} = \vec{F}/m$ that the net force is zero. Since the net force due to the two forces is zero, the forces must be equal in magnitude and oppositely directed.
14. The floor exerts an upward contact force, $\vec{F}_n$, on the soles of your feet. The reaction force is the downward push of the soles of your feet on the floor.
15. True. Each object has its own coordinate system, so an acceleration in the $+y$ direction in the

coordinate system for object one could correspond to an acceleration in the $-y$ direction of the coordinate system for object two. Usually it is much less confusing to avoid this situation by choosing your coordinate systems so that both objects accelerate in a positive direction.

16. If you consider a small piece of string with mass m, there is a tension T_1 pulling it to one side and another tension T_2 pulling it to the other side. According to Newton's second law, $T_1 - T_2 = ma$. (There is a minus sign because the tensions act in opposite directions). If the mass is negligible, then the product ma is always zero, so $T_1 - T_2 = 0$. The only way this can happen is if the magnitude of the tension throughout the string is uniform.

TRY IT YOURSELF–TAKING IT FURTHER

1. You cannot say for sure if simply pushing with more force will help your friend without doing rigorous calculations. Why? If you push with more force, the x component of your force will certainly go up. This means the x component of the force from your friend can go down. However, the y component of your force will also get larger in the negative direction, so your friend might have to pull up with more force. In addition, as you will learn in the next sections, the vertical force of the floor and the force of friction also depend on the vertical component of your pushing force. To guarantee that you will help your friend out, simply push the box in a more horizontal direction. This will increase the horizontal component of your force, which will reduce the horizontal component of the force your friend must exert. In addition, it reduces the downward force you exert on the box, which will reduce the frictional force, and the upward force your friend needs to exert.

2. The three forces acting on the puck are the tension of the string, the force due to gravity pulling the puck down, and the force of the ice pushing up on the hockey puck.

3. The acceleration of the object will be the same, because its mass is constant. However, its weight will be approximately one-sixth of that on Earth because the gravitational force on the moon is smaller.

4. In order to stretch the spring, the tension in the rope would have to be larger than the weight of the box.

5. The sole advantage of the one-wire system is its simplicity—only one wire is needed. The two-wire system has several advantages, even though two wires are required. First, the wire does not necessarily need to be as strong. In the one-wire system, the single wire must support the entire weight of the traffic light. With the two-wire system, the wires can share support duties. Second, supplying two wires as shown makes the light much less vulnerable to winds along the horizontal axis. The light may pivot about the attachment point, but as long as the wires hold, it will not move sideways.

6. The acceleration will be reduced, because the net force in the x direction will be reduced. The normal force of the table on the block will also be reduced because the y component of the tension in the string will also act against the block's weight.

7. The minimum tension would be with T_2 straight horizontal. In this situation, the magnitude of T_2 will have to cancel the x component of T_1, and the y component of T_1 will be equal in magnitude to the weight of the ball. So $T_2 = 35.7$ N.

8. No, the problem would not change at all. The source of the tension T_2 is irrelevant.

9. This answer is different because in this problem box A is "trailing" box B, which is the opposite of the situation described in the previous problem. Box A is trailing in the sense that the external force causing the acceleration acts directly on box B and is only transferred to box A via T_1.

QUIZ

1. True
2. False
3. Your inertia carries you forward. Only the force of the seat on your bottom holds you back.
4. Measure and compare the initial accelerations caused by each force on a mass m.
5. This statement is false. These are third-law pairs, and as such they must have the same magnitude. The pony and the cart accelerate forward because the pony pushes backward on Earth.
6. (a) $a = F/(m_1 + m_2)$; $F_c = F[m_2/(m_1 + m_2)]$; (b) $a = 0.400$ m/s^2; $F_c = 2.40$ N
7. $2m_1 m_2 g/(m_1 + m_2)$

Chapter 5

COMMON PITFALLS

1. False. It is only the *maximum* static frictional force that is proportional to the normal force.
2. The floor applies a frictional force to the soles of your shoes in response to your shoes pushing the floor. This force is directed in the "forward" direction.
3. True. Since the drag force increases with speed, eventually the upward drag force will equal the downward force of gravity on an object.

4. No. The drag force is proportional to v^n. The drag force certainly slows you down. Assuming the drag force could bring you to a complete stop, your velocity would be zero, and the drag force would vanish, making it impossible for it to reverse your direction.

5. True. It is possible for almost any type of force acting on an object to have some component pointing toward the center of curvature of the motion.

6. One hopes it is static friction; if it were kinetic, the tires would skid on the surface. The centripetal component occurs when you turn the tires into the turn.

7. False. Internal forces necessarily come in action–reaction pairs, so they always sum to zero. The acceleration of the center of mass depends only on the net *external* force on the system.

8. The center of mass is not necessarily part of the object itself. Consider a disk with a mass m uniformly distributed throughout it. Clearly the center of mass in this situation is the center of the disk. Now imagine removing material from the center of the disk out to some radius r, leaving you with a hole in the center. The center of the remaining ring is still the center point of all the mass, even though there is no material there.

TRY IT YOURSELF–TAKING IT FURTHER

1. (a) The frictional force will remain the same. As long as the objects are sliding, frictional forces are roughly constant over a very large range of velocities. (b) No. The tension will increase by a smaller amount. (c) If the tension also increased by 10.0 N, then the net horizontal force on the top block would still be zero, so the blocks wouldn't accelerate.

2. Once the car starts to skid, kinetic friction takes over. Since the coefficient of kinetic friction is less than the coefficient of static friction, the frictional force will be reduced from the maximum possible stopping power, so the magnitude of the car's acceleration will be reduced, and the car will travel a greater distance.

3. No. Because the frictional forces we have talked about are independent of speed, the only thing a frictional force can accomplish is to reduce an object's acceleration. Over the same time interval, this will reduce the speed an object can achieve, but it does not fundamentally limit the speed that object can obtain.

4. The tension will be the largest at the bottom of the circle, because at that point the tension must both counter gravity and provide a net upward force to keep the yo-yo moving in a circle. The tension will be the least at the top of the circle, because the gravitational force is centripetally directed, so the string does not have to provide as much centripetal force.

5. The roller coaster must travel at speeds greater than the speed you found. At higher speeds, the centripetal acceleration is larger, so a greater centripetal force is required. This force comes from the seat of the roller coaster.

6. Yes. Any symmetric distribution of the mass will have this property. Other more complicated mass distributions could also be constructed so the center of mass of the rod is located at its midpoint.

7. The force of the peg will increase. As the weight moves closer to the peg, the center of mass also moves closer to the peg, which makes the radius of the motion smaller. Since $F \propto 1/r$, the force will increase.

QUIZ

1. True. If a thrown grenade explodes in flight, after the explosion no fragments remain near its center of mass. Consequently, it is possible for an object to go unscathed even though the center of mass passes through it.

2. False

3. According to Newton's third law, when the foil deflects the air upward the air will push the foil downward. This increases the normal force of the car on the road, which is desirable because the maximum frictional force is proportional to the normal force, and it is this frictional force that provides the necessary centripetal acceleration.

4. Curved roads are banked so the normal force of the road on the tires has a centripetal component. This component supplements the frictional force; that way friction alone does not have to provide the necessary centripetal force, and there is less danger of skidding off the curve.

5. Yes. There is nothing special about this number.

6. $\vec{r}_{cm} = (1.00 \text{ m})\hat{i} + (1.00 \text{ m})\hat{j}$

7. $T = 1.14$ N

Chapter 6

COMMON PITFALLS

1. False. Work is negative when the force and displacement are oppositely directed.

2. The total work done on the block equals its change in kinetic energy. Its initial kinetic energy is zero and its final kinetic energy is $\frac{1}{2}mv_f^2$, where m is its mass and v_f is its velocity as it reaches the bottom of the incline. Thus the total work is $\frac{1}{2}mv_f^2$. Because the normal force is perpendicular to the displacement, the work done by it is zero. The only other force acting on the block is the gravitational force. The work W_g done by this force must be equal to the total work done on the block. That is, $W_g = \frac{1}{2}mv_f^2$.

3. True. Otherwise the change in kinetic energy cannot equal the work done. The dimensions are ML^2/T^2.

4. The total work done on the ball equals its change in kinetic energy. Its final kinetic energy is zero and its initial kinetic energy is $\frac{1}{2}mv_0^2$, where m is its mass and v_0 is its velocity at its lowest point. Thus the total work done on the ball is $-\frac{1}{2}mv_0^2$. Two forces act on the ball: the gravitational force and the tension force exerted by the string. The work done by the tension force is zero because the ball's velocity, and therefore the infinitesimal increments of its displacement, is always perpendicular to this centripetal force. Even though the angle between the gravitational force and the displacement of the ball varies, we still know that the total work on the ball must have been done by the gravitational force: $W_g = -\frac{1}{2}mv_0^2$.

5. False. The dot product results in a scalar quantity.

6. $P = \vec{F} \cdot \vec{v}$. The tension force of a pendulum is always directed centripetally. The velocity of the mass is always in a tangential direction. Since these vectors are perpendicular, their dot product, and hence the power delivered by the tension force, is always zero. However, the force of gravity also acts on the pendulum, and it is this force that is responsible for the change in kinetic energy of the pendulum.

7. False. The statement is true only if the points of application of each force undergo equal displacements. Although this condition must hold for a particle, it does not necessarily hold for an object. For example, if you compress a spring by pushing on opposite ends with your hands the two ends do not move through identical displacements. (Even if they are identical in magnitude they will not be identical in direction.)

8. The work–kinetic-energy theorem relates the total work done on a particle with the change in the particle's kinetic energy. The total work done on the book equals the change in kinetic energy of the book. The initial speed of the book is zero and the final speed of the book is zero, thus the change in kinetic energy is zero minus zero which equals zero. The total work done on the book is zero. The gravitational force does negative work on the book, but your hand does positive work. The work–kinetic-energy theorem tells us that the sum of the two works must equal zero.

Try It Yourself–Taking It Further

1. No, it won't. If the applied force $\vec{F}$ is suddenly removed, the box will continue sliding up the ramp for some distance, while the force of gravity does negative work and slows down the box, eventually stopping it. From that point, the force of gravity will do positive work on the box, constantly accelerating the box as it slides back down to its original position. The box's speed at that point could actually be larger than its speed when the applied force $\vec{F}$ is turned off.

2. Try to align the force $\vec{F}$ with the angle of the ramp, so it is pushing in the direction of the block's displacement.

3. Its speed would be 1.00 m/s. The area under the curve in this region adds to zero, so no net work is done, which means the particle's kinetic energy, and hence speed, must be the same at the start and end points.

4. Yes. The spring does negative work, and the applied force does positive work. Because the work done by the spring depends on x^3, and the work done by the applied force depends on x, eventually the negative work of the spring will equal the positive work done by the applied force. At the value of x when the two works sum to zero, the speed of the box will be zero.

5. No. We can use the work–kinetic-energy theorem only if we know the initial speed. Without that information, we can only determine the *change* in the particle's speed.

6. In these examples the normal force is always perpendicular to the object's velocity, so it's power is always zero. However, because objects can move up or down ramps which are not perpendicular to the gravitational force, this force *can* deliver power. If an object slides down a ramp, or experiences freefall, the gravitational force will deliver positive power to the object.

7. We also have to integrate F_y to find the work done by that component of the force. This would have the effect of increasing the speed of the particle even further. We would also have to take into account F_y when calculating the final power.

8. The total work of me on the weights would still be the same. I would do positive work in lifting the weights up, but an equal amount of negative work when lowering the weights. It feels like more "work" is required because our muscles are twitching for longer periods, so more energy is required to

change our muscle's kinetic energy.

1. False
2. True
3. Because the structure of the car does not remain fixed as the car moves, the car cannot be modeled as a particle. This means the work–kinetic-energy theorem does not apply. Neglecting air resistance, there are two forces acting on the car, the force of gravity and the contact force of the road on the tires. The force of gravity does work $-mgh$ where m is the mass of the car and h is the altitude gained by the car. The work done by the contact force of the road on the car is zero because the displacement of the point of application of this contact force is zero. (Unless the car is skidding, the part of the tire in contact with the road is not moving.) Thus the total work done on the car equals $-mgh$. If air drag were not negligible it would do additional negative work on the car. The car's speed, and hence kinetic energy, does not change because the energy stored in the fuel of the car gets converted to propel the car forward. This is a change in the internal structure of the car, so once again, the work–kinetic-energy theorem does not apply.
4. Two forces act on you when you start walking: the force of gravity and the contact force of the floor on your feet. Assuming that you are walking on a horizontal floor, the force of gravity does no work because it acts at right angles to your displacement. The contact force of the floor only acts on a foot when that foot is in contact with the floor. Thus, whenever this force acts on it, the foot is at rest so the displacement of the point of application of the contact force is zero. Therefore the contact force does no work. Because neither force acting on you does any work, the total work done on you is zero. Your increased kinetic energy comes from the chemical potential energy that is stored in your muscles. These chemical changes constitute changes in your internal structure. An object whose structure changes cannot be modeled as a particle, but the work-kinetic energy theorem only holds for particles. For this reason this theorem does not apply to your motion. This type of problem is analyzed using another theorem, the work-energy theorem, which is presented in Chapter 7.
5. Some component of the force must be directed in the opposite direction of the particle's displacement.
6. (a) 3.00 J, (b) +9.00 J, (c) 4.36 m/s
7. 50 kW

Chapter 7

1. False. They have more potential energy when they are far apart because $\Delta U = -W$, and as the objects are separated, the work done is negative, so the potential energy increases.
2. The potential energy of a spring is $U = \frac{1}{2}kx^2$, where x is the position of the end of the spring relative to the equilibrium position. This is an even function, so it does not matter if x is positive or negative. In either situation, if you let go of the spring the potential energy will be converted into kinetic energy.
3. True
4. Zero. There are no external forces acting on the block–slide–earth system, so the total energy of the system remains constant. The slide is frictionless so no mechanical energy is dissipated by friction. The only force doing work on the block is the conservative force of gravity. Therefore, for this system mechanical energy is conserved. The decrease in gravitational potential energy equals the increase in kinetic energy.
5. False. Some of the dissipated kinetic energy appears as the thermal energy of the block, almost all the rest as the thermal energy of the floor.
6. The total energy of this system increases. As the boy pulls the wagon up the hill, the force he exerts on the wagon does positive work. This is the only external work done on the system, so the total external work done on the system is positive. This means the system's total energy increases, in accord with the work-energy theorem. Where does this energy come from? The chemical energy stored in the boy's muscles.

1. For both the x and y cases, the $d\vec{\ell}$ changes direction on opposite sides. However, for the vertical segments, the work integral is $-3x^2y$. The $-3x^2$ portion has the same sign for both the left- and right-hand segments. For the horizontal segments the work integral is $2yx^2$. On the top segment $2y$ is positive, but on the bottom segment $2y$ is negative. This has the effect of canceling out the negative sign from $d\vec{\ell}$, so the result is another positive work.
2. If the particle is attached to the spring, then by the time the spring has stretched a distance d, it will have done enough negative work on the particle so that its speed will be zero. If the particle is

not attached, the spring cannot "pull" on it to slow it down. As a result, once the spring reaches its equilibrium position, the mass will continue to move with a speed of v_f.

3. Power is $P = \vec{F} \cdot \vec{v}$, so we need both a large force and a large velocity. This will occur somewhere along the steepest parts of the potential-energy graph. Whether the larger velocity occurs near the top or bottom of those regions depends on the initial position and velocity of the particle.

4. No. The gravitational force does a total amount of work equal to $+m_1 gh$ on mass 1, and $-m_2 gh_2$ on the second mass.

5. The spring would compress less. More of the system's initial mechanical potential energy would be converted to heat, so less potential energy would have to be stored by the spring.

6. The coefficient of kinetic friction depends on *both* materials that are in contact. The sled–ice combination can have one set of coefficients of static and kinetic friction while the paw–ice combination can have different coefficients of static and kinetic friction.

QUIZ

1. False. The displacement of the point of application of the frictional force is not equal to the bulk displacement of the object that the force acts on.

2. True

3. The total energy of this system does not change. There are no external forces acting on this system, so its total energy remains unchanged.

4. No external force does work on the car. The static frictional force on the tires by the pavement does no work because that part of each tire in contact with the pavement is at rest. (We are assuming the tires do not slip on the pavement.) The increase in the car's kinetic energy equals the decrease in the chemical energy stored in the gasoline and the oxygen in the atmosphere. The kinetic energy comes from the gasoline-oxygen mixture.

5. No. The total energy of the putty–floor–earth system does not change because there are no external forces on the system, so no work is done on it by external forces. During the fall the system's gravitational potential energy is transformed into the kinetic energy of the blob and during the plop this kinetic energy is dissipated via friction within the putty and between the putty and the floor.

6. 1.5 m

7. $b = 0.4$. Negative tension indicates that the particle does not actually reach the height bL above the peg, because the the string goes slack.

Chapter 8

COMMON PITFALLS

1. False. Internal forces necessarily come in action–reaction pairs, so they always sum to zero. The rate of change of the momentum of the system equals the total *external* force on the system. External forces also come in action–reaction pairs, but the force of the system on the external agent is irrelevant to the system's momentum.

2. The momentum of the ship–fuel–rocket system remains constant, as there is no net external force on this system. The momentum of both the ship and the rocket do change, because each exerts a net force on the other. Any changes in the momentum of the ship are balanced by oppositely directed changes in the momentum of the rocket.

3. False. The velocity of the center of mass, along with the associated kinetic energy, is the same after the collision as before. If it is nonzero before the collision it is nonzero following the collision.

4. The impulse required to cause a specified change in momentum is fixed, thus the area under the force-versus-time curve is fixed. The longer the time interval, the smaller the maximum force required. The force must be at least as large as the average force at some time in the interval. Longer times mean a smaller average force, which means a smaller maximum force is possible.

5. False. There is still kinetic energy due to the motion of the objects relative to the center of mass.

6. Yes, it is. In a reference frame attached to the object, the object's speed appears to be zero, and the object's surroundings appear to be moving.

TRY IT YOURSELF–TAKING IT FURTHER

1. If the butt of the rifle is not well-supported, it will travel backward with a pretty good speed, and the resulting impact will likely hurt someone or something.

2. Momentum will still be conserved in both the x and y directions. However, the car and truck will not be traveling in the same direction. This has the effect of introducing two additional variables into the problem. As a result, the problem cannot be solved without additional information.

3. This collision is partially inelastic. We can tell by calculating the initial and final kinetic energy of the system. The initial energy is 64 J, and the final energy is 35.3 J. Because energy is lost, this is an

inelastic collision.

4. Yes. If the railroad worker throws the water off the back of the train car she will at the very least reduce the forward momentum of the water, which will cause the train car to speed up to compensate. Whether or not the train car reaches or exceeds its initial velocity depends on the impulse the worker can impart to the water.

5. The velocities of each object still get simply reversed relative to the center of mass. However, the calculation of the location of the center of mass and its velocity becomes a bit more complicated. In the laboratory frame, the cue ball will no longer stop after the collision.

6. To keep the comet traveling at a constant speed, there would have to be an external force acting in the direction of motion to counteract the $-Rv$ term of force imparted by the cosmic dust.

QUIZ

1. True

2. False

3. When a system has zero kinetic energy, nothing is moving. (There are no negative terms in the kinetic-energy sum.) Thus the momentum must be zero. When a system has zero momentum, the center of mass is at rest. The system can still have kinetic energy due to the motion of parts of the system relative to the center of mass. A spinning ball necessarily has kinetic energy, but may have zero linear momentum.

4. If these forces were not equal in magnitude and oppositely directed, the impulses imparted during the collision, and hence the changes in momentum of the car and the truck, would not be equal in magnitude and oppositely directed, and the momentum of the car–truck system would not remain constant.

5. No. Momentum is conserved only in directions in which the net external force is zero. It is conceivable for a "gentle" collision to not conserve momentum at all, if the external forces are very large.

6. $F = 2.05 \times 10^{-2}$ N

7. $\vec{v}_{2 \text{ ball}} = 0.5v_0$ at a direction of 60° from the original trajectory.

Chapter 9

COMMON PITFALLS

1. True

2. (a) Both points have the same angular acceleration. This, of course, is the point of describing rotational motion in terms of angular quantities. (b) The tangential acceleration $a_t = r\alpha$ of the point on the rim is larger because it is farther from the axis. (c) The radial acceleration $a_{\text{rad}} = v^2/r = r\omega^2$ of the point at the rim is larger for the same reason. (d) Since the radial acceleration and centripetal acceleration are the same, the answer is the same as for (c).

   Note: The centripetal direction is always defined as radially inward (toward the rotation axis). The radial direction is less definite. Sometimes radially inward is taken as the positive radial direction, but more often, radially outward is considered positive. Thus, $a_{\text{rad}} = \pm a_t$ depending on whether radially inward or outward is taken as the positive direction.

3. True. The moment of inertia depends only upon the mass of the object and how it is distributed relative to the axis.

4. Yes. The moment of inertia may be different about different rotational axes.

5. False. Consider two forces of equal magnitude and opposite direction. The forces add to zero, but the net torque is zero only if they act along the same line of action.

6. No. The gravitational torque on the system decreases. The weight force remains constant but its lever arm ℓ decreases as he swings down toward the bottom. This decrease in torque results in a decrease in the angular acceleration. The constant-angular-acceleration formulas are valid only if the angular acceleration remains constant.

7. False. The centripetal acceleration of any part of the object is proportional to its distance from the axis of rotation.

8. Certainly. You are free to choose any coordinate system you would like for each object separately. However, it is usually easiest to choose your coordinate systems so that all object experience a positive acceleration in their respective coordinate systems.

9. False. Consider a wheel, whose rotation axis is the contact point between the wheel and the road, which is always moving. Furthermore, the axis of rotation can change direction, as when a bicycle wheel turns a corner.

10. No, it can slip. Consider the spinning of car tires stuck in a snowbank.

Try It Yourself–Taking It Further

1. It corresponds to a constant *tangential* acceleration for all points of the turntable that are the same distance from the central rotation axis. Points farther from the axis will experience a larger a_t than points closer to the axis. In this problem the centripetal acceleration for all points will decrease as time goes by since $a_c = r\omega^2$.

2. The frictional force keeps the dime accelerating toward the center and in the tangential direction. Once the frictional force can no longer apply the required force to keep the dime stationary with respect to the turntable, the dime will then slide both outward, and "backward" relative to point P.

3. This is a complicated question, with several inputs. First, because the dime is closer the turntable must achieve a greater angular acceleration before the dime will slip. This means the kinetic energy of the turntable will increase. However, the tangential speed of the dime is reduced, so its kinetic energy goes down. Which of these factors dominates depends on the new location of the dime.

4. It will be easiest to spin the structure around the axis with the smallest moment of inertia: the x axis.

5. The smallest moment of inertia will be around an axis that symmetrically divides the triangle. This would be an axis that goes through the points $(0, \frac{a}{2})$ and $(b, 0)$.

6. The torque would decrease because the lever arm would be shorter.

7. More force is required. $\tau = F_t r$, so if r is reduced, then F_t must increase to compensate.

8. T_2 must be larger than T_1 in order for the pulley to have a clockwise angular acceleration. There is no contradiction, because m_1 is smaller than m_2.

Quiz

1. False

2. False

3. The net torque about the helicopter's center of mass must equal zero, otherwise it will start to rotate. The rotational motion of the main rotor is counterclockwise. Thus the drag force of air, which opposes this motion, produces a clockwise torque on the helicopter. To counter this clockwise torque, there must be a counterclockwise torque of equal magnitude acting on the helicopter. This is accomplished if the tail rotor pushes air to the helicopter's left (port). According to Newton's third law, if the tail rotor pushes air toward the left, then the air pushes the tail rotor to the helicopter's right (starboard), thus producing a counterclockwise torque on the helicopter. (The tail rotor's axis is oriented both horizontally and at right angles to the helicopter's forward direction.)

4. It is helpful to separately consider the translational and rotational kinetic energies. All the objects have the same linear speed so they all have the same translational kinetic energy. Also, they all have the same angular speed so the larger the moment of inertia, the larger the rotational kinetic energy. The hoop has the largest moment of inertia, the disk the next largest, and the sphere the smallest, so the hoop has the largest rotational kinetic energy and the sphere has the least. Each will roll up the hill until all its kinetic energy is transformed into gravitational potential energy, so the greater the kinetic energy, the higher up the hill it will roll. This means the hoop rolls the highest, with the disk second, and the sphere last. Of course, if all three start with the same kinetic energy instead of the same speed, they will all roll up to the same height.

5. True, from the parallel-axis theorem.

6. $I = (19/32)M_1 R^2$

7. $7v_0^2/(10g)$

Chapter 10

Common Pitfalls

1. True, from the definition of torque and the cross product.

2. No. When we change the order of the vectors being crossed, we end up with the opposite of the original answer. We need to get the same answer for the cross product to be commutative.

3. True

4. The tendency for the plane to turn sideways (yaw) is a gyroscopic effect. The propellers of single engine aircraft of U.S. manufacture rotate clockwise as seen from the rear. In accord with the right-hand rule, the angular momentum vector $\vec{L}$ of the propeller points in the forward direction, parallel to the propeller shaft. The downward force of the air on the tail of the plane results in a vector torque about the center of mass in the starboard direction. This causes the plane to precess, slowly turning to starboard so that $\Delta \vec{L}$ is in the same direction as the net torque. To prevent this rotation the pilot uses the rudder to deflect the air to the port. The air then exerts a starboard force on the rudder, which tends to turn the nose of the plane to port.

5. False. A pair of non-collinear (not acting along the same line) forces that are equal in magnitude but

opposite in direction produce a net torque but not a net force. These forces will change the system's angular momentum, but not its linear momentum.

6. The ball does lose angular momentum during the collision. Its angular momentum is given by $\vec{L} = \vec{r} \times m\vec{v}$, where $\vec{r}$ is the vector from the door's rotation axis to the ball, m is the ball's mass, and $\vec{v}$ is the velocity of its center of mass. During the collision the magnitude of $\vec{v}$ decreases so the angular momentum of the ball decreases. This decrease equals the increase in the door's angular momentum.

TRY IT YOURSELF–TAKING IT FURTHER

1. No. In order for circular motion to be occurring about that point, $\vec{r}$ and $\vec{F}$ would have to be perpendicular, which they are not.

2. $\vec{C} = \vec{B}$. By definition the cross product is perpendicular to the vectors being crossed, so it must also be perpendicular to $\vec{C}$.

3. The value of the torque, and hence the change in angular momentum would change, but $\vec{\tau}$ would still be equal to $d\vec{L}/dt$ about the new origin.

4. The merry-go-round would still acquire an angular acceleration, but a smaller one, because the angle between $\vec{r}$ and $\vec{F}$ will be reduced. If the rabbit jumps radially, then the merry-go-round will not begin to rotate.

5. The boy is running in the direction of rotation of the merry-go-round. The angular speed, and hence angular momentum of the merry-go-round decrease. The boy's angular momentum must increase, *in the same direction*, if angular momentum is to be conserved.

6. No, he can't. The required speed of 19.0 m/s is nearly twice that of the fastest person in the world. He will need some help to make it all the way over the Ferris wheel.

QUIZ

1. False

2. False.

3. No, it will not, because there is no net gravitational torque.

4. No. The stated criterion only guarantees static balance, not dynamic balance.

5. Spin isn't a force. For a spinning gyroscope supported at a point, it is the force of the support exerted on the gyroscope that prevents it from falling. The force of the support is equal in magnitude and opposite in direction to the gravitational force. However, they do not necessarily act along the same line unless the gyroscope is supported at a point along the vertical line through its center of mass. The gyroscope precesses as a result of the torque due to the gravitational force.

6. 3.62 m/s

7. 3.0×10^{-4} s

Chapter R

COMMON PITFALLS

1. False. Since both events take place at the same location in the reference frame of the first observer, 2 s is the proper time interval between them. Any observer moving relative to the first observer will find a time interval larger than 2 s because of time dilation.

2. Yes. The lasting effects of the simultaneous events must be present in all reference frames.

3. False. Both observers will agree about lengths along the y axis.

4. If the object is moving relative to you, you measure its length by finding the difference between the coordinates of its end points at the same time.

5. True

6. Yes. The time ordering between the two events as determined in reference frame S′ can be positive, zero, or negative depending on the spatial separation, the time difference $T_A - T_B$ in reference frame S, and the relative velocity v.

7. True

8. When the total energy $E \gg mc^2$, or equivalently when the kinetic energy $K \gg mc^2$, you can use $p = E/c$ to a good approximation. This is true even for photons, whose mass is zero.

TRY IT YOURSELF–TAKING IT FURTHER

1. The observed time by the track must be longer. It will be 1.67 ms.

2. She also thinks she is traveling at 0.8c. To confirm this, divide the contracted length she measures by the time she measures. The relative speed of two inertial coordinate systems must be the same regardless of the coordinate system in which it is measured.

3. If the given energy is the total energy, the problem wouldn't be possible. The proton has a rest mass of 938 MeV, so the energy given is not enough even for the proton to exist. If the given energy is

the kinetic energy only of the proton, then you can use nonrelativistic expressions because the given energy is significantly smaller than the proton's rest energy.

4. The rest mass of a neutron is only slightly larger than the rest mass of the proton, so relativistic expressions will still be required. The kinetic energy of the neutron will be slightly less because its rest mass is slightly larger than the proton's.

QUIZ

1. True
2. True
3. The observer in S also measures the light signal to move with speed c.
4. Since the red and blue flashes are both emitted at the same place as determined in reference frame S′, the time interval of 5 s determined by a single clock in S′ is the proper time interval between the two events.
5. Yes
6. $v = 0.99999956c$
7. $K = 300$ MeV

Chapter 11

COMMON PITFALLS

1. False. The law of equal areas follows directly from the conservation of angular momentum. It holds for any and all central forces, not just those that vary inversely with the square of the distance.
2. An ellipse is the locus of points for which $r_1 + r_2 =$ constant. You can draw an ellipse with a piece of string by fixing each end at a focus and using it to guide the pencil as you stretch the string taut around the ellipse. Because the string doesn't change length, the sum of the distances from each focus of every point will remain constant.
3. False. The ratio of the moon's acceleration to the acceleration of objects falling freely at the surface of Earth is R_E^2/r^2, where R_E is the radius of Earth and r is the radius of the moon's orbit. Measurement of this ratio is the confirmation of Newton's law of gravity.
4. From Kepler's second law we know the line joining Earth and the Sun sweeps out equal areas in equal times. In January, when this line is shortest, it sweeps out angles the fastest. It follows that viewed in January from Earth, the Sun's apparent motion against the background of the "fixed" stars is greater than it is in July. This can be observed by comparing the sidereal day and the solar day. (The sidereal day is the time for each "fixed" star to reach its apex on consecutive days, whereas the solar day is the time for the Sun to reach its apex on consecutive days. The apex is the highest point reached.)
5. True
6. No. The zero point of the potential energy is somewhat arbitrary. However, when using Newton's law of gravitation, it is generally most convenient for the zero point to be infinitely far from the object, and to have the gravitational potential energy be negative everywhere else in space.
7. True. The $\vec{r}$ points away from the mass, but because of the negative sign in the expression for the field, the gravitational field itself points toward the object creating it.
8. One mass. That is it. If a mass exists, it creates a gravitational field throughout all of space. A second mass is not required.

TRY IT YOURSELF–TAKING IT FURTHER

1. The angular momentum is given by $\vec{L} = \vec{r} \times \vec{p}$. Since we are approximating Saturn's orbit as circular, this reduces to $L = rp$. The mass of Saturn is approximately 5.68×10^{26} kg, so $L = 7.78 \times 10^{42}$ kg·m^2/s.
2. Assuming a circular orbit, the central force of gravitation is providing the acceleration of the planet: $GM_S m_J/r^2 = m_J v^2/r$. From this expression, you should convince yourself that smaller orbital radii require higher orbital speeds. Since the orbital radius of Jupiter is less than that of Saturn, its orbital speed should be higher.
3. The planet is less massive and has a larger diameter than does Earth. As a result, the local acceleration due to gravity, $g = GM/r^2$, will be smaller than that for Earth, so you will feel substantially lighter.
4. The weight of a fly is approximately 3.92×10^{-3} N. We do not notice the increased weight when a fly lands on our hand. So we certainly will not notice, or be drastically affected by, the gravitational forces in this problem, which are 1000× smaller than that!
5. The satellite would simply fall back down to Earth. Without any tangential speed, the satellite cannot maintain its orbit.
6. Our answers agree with Kepler's second law. To sweep out the same area when the satellite is close, it must travel a greater length of arc within a given time as compared to when the satellite is distant. This can only be accomplished if $v_p > v_a$, which is the situation here.

7. No. There is no symmetry at this point to cause the gravitational field to become zero.

8. No. It is only the specific location of point P relative to the length of the mass that allows us to arrive at a numerical value for the gravitational field.

QUIZ

1. True: period^2 $\propto$ radius^3.

2. False. It is the force per unit mass, a vector.

3. Properly speaking, the astronaut isn't weightless since weight is the force that gravity exerts on him or her. Earth's gravity hasn't gone away. However, both the astronaut and scale are in free fall, so the scale exerts no net force against the astronaut and his or her apparent weight is zero.

4. There is more residual atmosphere at lower altitudes. Thus, most of the effect of atmospheric drag on the satellite occurs when it is near perigee (its point of closest approach to Earth). Every time it comes around to perigee, it loses a little speed. As a result, it doesn't climb as far away from Earth during the next orbit, so its path becomes a little more nearly circular each time.

5. If a satellite is in a circular equatorial orbit with a period equal to that of Earth's rotation, it will appear to be stationary relative to a single spot on the ground. Such an orbit is called a geosynchronous orbit. The altitude of the required orbit, which can be found using Kepler's third law ($T^2/r^3 = $ constant) using the satellite's orbital data and the radius of Earth, works out to about 35 800 km, or 22 400 mi.

6. $v = 1950$ m/s. $K = 2.43 \times 10^{10}$ J

7. $h = 7.71 \times 10^5$ m

Chapter 12

COMMON PITFALLS

1. False. The torque exerted by a couple about a point P is always independent of the location of P.

2. No. For example, the center of gravity of a bowl is outside the material of the bowl.

3. True. They have dimension of $M/(LT^2)$ and SI units of N/m^2.

4. Steel has relatively high tensile and compressive strength, as well as a high shear modulus as compared to many other materials. Practically, this means steel can support large weights set on top of it, support large weights hung from it, and also withstand shear or torsional forces much better than most other materials. It will function well (not be prone to deformation) regardless of the geometric configuration in which it is placed.

TRY IT YOURSELF–TAKING IT FURTHER

1. Since the girl is sitting much further from the pivot, in order for the torques to be zero her weight, and hence mass, must be less than that of the boy. The plank also exerts a torque in the same direction as the torque from the girl's weight, further reducing the required mass of the girl.

2. No. The top string will always exert some force to the left, which must be countered by some force to the right from the hinge. It may be possible to eliminate the y component of the force from the hinge, however.

3. If normal steel is used, then the steel will break first, because it has a lower tensile strength than tungsten. If high-tensile steel is used, then the tungsten will likely break first.

4. The shear angle will increase because the same force is distributed over a smaller area.

QUIZ

1. True

2. False. The lever arm equals the perpendicular distance from P to the line of action of the force.

3. Yes. If a point is on the line of action of a force, then the force's torque about that point must equal zero. Conversely, if the torque of a force about a point is zero, then the point must lie on the force's line of action. (Convince yourself that these assertions are true; make some drawings.) The intersection of point P of the lines of action of two of the forces lies on both lines, so the torques of both forces about P each equal zero. Since the sum of all three torques about P must equal zero, and the torques of two of the forces are both zero, the torque of the third force must also equal zero. It follows that the line of action of the third force must also pass through P.

4. No. You can choose any rotation point you like. However, some points result in expressions that are algebraically simpler than those you might get by choosing less convenient rotation points.

5. The higher the sign, the greater the torque about the horizontal axis through the lowest extremes of the posts for a given wind speed. In addition, the deeper the post hole the greater the maximum opposing torque about the same axis for a given consistency of dirt. So higher signs require deeper holes for the sign posts.

6. $m = 153.2$ g, $x = 46.6$ cm from the lower-left end

7. $T = 110$ N, $\vec{F} = (110\hat{\imath} + 147\hat{\jmath})$ N

Chapter 13

COMMON PITFALLS

1. True. This is the definition of specific gravity.

2. Yes; consider a canoe that floats on water. Another possibility would be a solid object that floats on a fluid with a specific gravity greater than the object—the fluid is not specified here.

3. False. The pressure varies with depth because of the weight of the fluid lying above. However, any *change* in pressure at one point in a fluid results in an equal *change* at all points in it.

4. The fluid added the second time wasn't water but some liquid that is less dense than water and doesn't mix readily with it. Thus, a taller column of the second fluid was required to produce the same pressure in the liquid at the bottom of the U-tube.

5. True

6. When the ice is floating, the part of it that is underwater is displacing a volume of water that is equal in weight to the whole weight of the ice. This is exactly the volume of the water produced by the melted ice, so the water level goes neither up nor down but stays the same. (Note that frozen water shrinks on melting.)

7. False. A fluid cannot sustain a static shearing stress, but shearing stresses (called viscous forces) do exist in a real flowing fluid. They do not exist in an ideal fluid.

8. The pressure decreases because the fluid has viscosity. If a steady flow is to be maintained, there must be a pressure difference to maintain the flow of the fluid against this friction.

TRY IT YOURSELF–TAKING IT FURTHER

3. The change in volume of the steel block would be smaller because steel has a larger bulk modulus.

4. No. In both cases the measuring height is the same relative to your heart as well as the rest of your upper body, so the blood pressure shouldn't change by much.

5. Yes. Even if the oil slick was so thick the raft floated entirely in oil, the raft would still float. The specific gravity of oil is more than half that of water. So the raft has to displace less than twice the amount of oil to still float. That would mean it would sit less than 15 cm in the water, which is less than the thickness of the raft.

6. It would float lower in the water. This may seem counterintuitive; but if the oil is removed, it is effectively replaced by air, which has a much lower specific gravity. So to displace the same total mass, the wood will have to displace more water, by lowering itself.

7. Yes. The Reynolds number is much less than 2000.

8. In the summer. The viscosity is three times lower at $60°C$ than at freezing, so the volume flow rate is significantly higher in warm weather.

QUIZ

1. False. It is derived by applying the work-energy theorem to account for the *change* in mechanical energy.

2. True

3. Suppose your mass is 70 kg. For you to float, the buoyant force of the water must equal your weight. Thus by Archimedes' principle, you must displace 70 kg of water. A 70-kg mass of water has a volume of 70 L or 0.070 m^3. The submerged 95% of your body must therefore have a volume of 0.070 m^3, so the total volume of your body must be $0.070 \text{ m}^3/0.95 = 0.074 \text{ m}^3$.

4. A suction pump raises water by creating a pressure in the pipe that is lower than atmospheric pressure; it is the pressure of the atmosphere that raises the water. Since the absolute pressure at the pump cannot be less than zero, the largest pressure available to raise the water is equal to atmospheric pressure, which corresponds to a height of about 34 ft of water.

5. To draw air into your lungs, your diaphragm must lower the pressure in your lungs to less than atmospheric pressure. At a depth of a foot, the water outside your body is already at a pressure around 3 kPa higher than atmospheric. Because 4 to 7 kPa is the maximum external pressure difference your diaphragm can handle, you can't suck air through a tube at a depth of much more than a foot.

6. $r = 4.33 \text{ m}$

7. 7.42 hours.

Chapter 14

COMMON PITFALLS

1. True

2. The particle's net displacement in one period is, of course, zero. The total distance it covers, however, is from $z = 0$ (its initial position) to $z = A$, from there to $z = -A$, and from $z = -A$ back to its initial

position. It therefore covers a total distance of $4A$.

3. False. This is its total mechanical energy; its kinetic energy isn't constant.

4. When the mass is halved, the frequency of the oscillation increases, but the total energy, $\frac{1}{2}kA^2$, is unaffected. When the amplitude is halved, the total energy decreases to $\frac{1}{4}$ of its original value.

5. False. It is simple harmonic motion only if the amplitude is small.

6. If the string has mass, a small fraction of the total mass that is swinging is closer to the pivot than the pendulum length L, so the effective length is a little less than L. Consequently, the motion of the pendulum has a slightly shorter period than it would if the string were massless.

TRY IT YOURSELF–TAKING IT FURTHER

1. No. The phase constant is 1.27 rad. At time $t = 0$, this is the value the argument of the cosine function must have to give the displacement shown on the plot.

2. The period is $T = 10$ s, so the frequency is $f = 1/T = 0.1$ Hz.

3. The kinetic energy is 0.225 J, or $\frac{3}{4}$ of the total energy. This is because the potential energy is $U = \frac{1}{2}kx^2$, so if x is halved, the potential energy is quartered.

4. $K_{max} = \frac{1}{2}mv_{max}^2 = U_{max} = \frac{1}{2}kA^2$. If the amplitude is increased by a factor of 10, the total energy is increased by a factor of 100, and the maximum speed is increased by a factor of 10.

5. As with all mass-on-a-spring systems, the maximum potential energy occurs when the mass is at an extrema of its motion. There is no difference between the upper position and the lower position, as long as the equilibrium position is established as the zero coordinate.

6. As with all simple harmonic motion situations, the maximum acceleration occurs when the mass is located at a displacement equal to its amplitude.

7. The Q would increase, because the energy losses would go down. This would increase the resonant frequency, and decrease the period of the child's oscillation.

8. None. By definition, when critically damped no oscillations occur.

9.

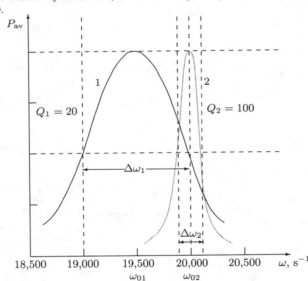

QUIZ

1. False. Damping increases the period.

2. True

3. What we have to find first is the force constant of half a spring. The same tension stretches one-half of the original spring half as much as it would the whole. Therefore, if the force constant of the whole spring is k, that of one of the halves is $2k$. Since the frequency is proportional to the square root of k, cutting the spring in half increases the frequency by $\sqrt{2}$.

4. The period of a simple pendulum is proportional to $\sqrt{L}$, so the period increases by a factor of $\sqrt{2}$ and the frequency decreases to $1/\sqrt{2}$ its original value. The energy of the swinging pendulum is not a constant because (negative) work is done on it by the hand that pays out the string. Since we don't know how to evaluate this work, we can't apply the conservation of energy to this problem. (We know the displacement of the hand, but the tension in the string is variable.)

5. It would have made no difference whatever. Defining simple harmonic motion in terms of the sine

does change the phase constant by 90°.

6. $T = 1.96$ s
7. (a) $v = 0.524$ m/s, (b) $\omega = 3.49$ rad/s, (c) $y(t) = (0.15$ m$)\cos((3.49$ rad/s$)t)$

Chapter 15

COMMON PITFALLS

1. False
2. You can generate longitudinal waves in a Slinky very easily by tying one end of the spring to a wall and stretching it to a moderate degree as shown in Figure (a). Moving the free end back and forth along the line of the spring will produce easily visible compression waves that travel at a few meters per second. Transverse waves are generated by moving the free end up and down as seen in Figure (b).

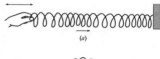

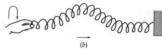

3. True
4. When the displacement is zero, the compression or expansion of the medium *around* it is maximum, which maximizes or minimizes the pressure at that point. However when the displacement is at a maximum or a minimum, then the molecules of the medium on either side are at the equilibrium separation, which results in the equilibrium, or zero, pressure. See Figure 15-11 on page 507 of the text.
5. False. The decibel is a unit of intensity level.
6. For a wave propagating uniformly in all directions in three dimensions, the intensity decreases inversely with the square of the distance from the source. The intensity is also proportional to the square of the amplitude. Thus the amplitude must decrease inversely with the first power of the distance from the source.
7. False. Reflections at a free end are *not* inverted.
8. The obstacles, apertures, and such that surround us in everyday circumstances—doorways, for instance—tend to be not very large compared to the wavelengths of audible sound, but all are very large compared to the wavelengths of visible light. Thus, sound waves are significantly diffracted, but light waves are not.
9. False. That the frequency received and the frequency of the source differ when there is relative motion between the source and the receiver is known as the Doppler effect.
10. You will hear the pitch rise and fall as the tuning fork is moved toward or away from you, but you will hear a nearly steady pitch as it is moved from side to side.

TRY IT YOURSELF—TAKING IT FURTHER

1. You could probably learn something, but not much. Both γ and M are properties of the air. You can learn something about their ratio, but other experiments are required to learn more about each quantity separately.
2. We know that energy depends on the frequency, which tells us that frequency is the fundamental quantity, not wavelength. The boy hears the same frequency, so the pitch does not change.
3. This is a transverse wave. It travels in the x direction, because x shows up in the argument of the sine function. The displacement is y, which is perpendicular to x.
4. With the help of Table 15-1 on page 512 of the text, we see that somewhere between 50 and 60 dB would probably be reasonable. This corresponds to an I of 10^{-7} to 10^{-6} W/m^2. You can back this out to find the required output power. It's not much!
5. The sound from the dogs is percussive. It usually consists of moments of silence separated by sharp barks. Traffic on a busy street, however, produces fairly steady, constant sound. Some people can use this "white noise" to actually help them fall asleep, whereas the sharp loud barks tend to startle a person.
6. If the pulse starts on the light string, it will be inverted upon reflection, and the reflection coefficient will be larger in magnitude than in this problem. The transmission coefficient will be less than 1.
7. The instant the car is passed, the observed frequency will change from 259 Hz to 242 Hz, because we will now need to change the signs used in the Doppler-effect formula. You have likely experienced this phenomenon as a car or train passes you while its horn is blowing.
8. The only way this is possible is for the driver to stop her car and drive with a speed greater than 40 km/h in the opposite direction before the sound wave has time to travel to the wall and back to the car.

1. True
2. True
3. No. In fact, just what "twice as loud" means isn't clear because loudness is a function of perception, but intensity level isn't. Typically, a tone sounds twice as loud when the intensity increases by a factor of 10, which corresponds to an intensity level increase of 10 dB. The 60 dB increase means the sound intensity is increased by a factor of 10^6 (or one million).
4. The diffraction of sound waves is the main reason you can hear "around corners" from the music room to the hall and then into your room. The smallest aperture involved is likely to be the width of a doorway, about 1 m. Sound waves with high frequencies and short wavelengths (much less than 1 m) diffract less than do waves with low frequencies and wavelengths much longer than 1 m. Thus, the high-frequency sound waves reaching your ears are diminished more than are the low-frequency sound waves. You will hear high notes, and particularly the high harmonics that give musical sounds their tone quality, less effectively, so the music will sound "flat" or "muffled" to you. Walls, drapes, and other surfaces selectively absorb some frequencies better than others. This will also affect what the listener hears.
5. Function (a) can be a wave function. Function (b) cannot be expressed as the sum of a function of $x + vt$ and $x - vt$, so cannot be a wave function. Function (c) is discontinuous, so cannot be a wave function. In textbooks, however, discontinuous functions are sometimes used as wave functions. You should think of these as continuous functions that are close approximations to the discontinuous functions.
6. $\lambda = 1.31$ m. $f = 269$ Hz
7. (a) $u_s = 39.2$ m/s, (b) $f_s = 389$ Hz

Chapter 16

1. False. The path lengths must differ by $(n + \frac{1}{2})\lambda$, where n is a positive integer and λ is the wavelength.
2. Figure (a) shows the displacement y of the string as a function of position x along the string; as the pulses approach each other, Figure (b) shows the velocity v_y of the string as a function of position x.

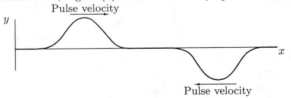

(a) Displacement of string as a function of position x along the string.

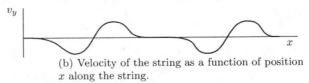

(b) Velocity of the string as a function of position x along the string.

When the pulses superpose, the resultant displacement and velocity of the string are the algebraic sums of the displacements and velocities of the individual pulses. At the moment the pulses completely superpose, the figure below shows the displacement y and velocity v_y of the string as functions of the position x along the string. At the moment the pulses exactly superpose, the energy carried by the two pulses is in the kinetic energy of the string.

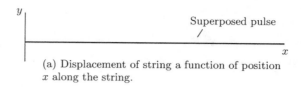

(a) Displacement of string a function of position x along the string.

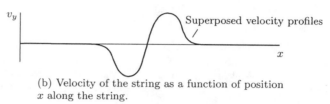

(b) Velocity of the string as a function of position x along the string.

3. False. At adjacent antinodes the motions are 180° out of phase. That is, when the displacement of the string at one antinode is $+A$ the displacement at the next antinode is $-A$.

4. The frequencies of the wind instruments increase because the speed of sound in air increases with increasing temperature. Since strings expand a little as the temperature goes up, the tension in a string decreases somewhat; the wave speed and fundamental frequency therefore decrease. Thus, wind instruments go sharp and string instruments go flat as they warm up. This is why an orchestra warms up first and then tunes.

5. True

6. In a way we do hear them. In this case the beat or difference frequencies are tens to hundreds of hertz, and we can't hear them as intensity variations. However, because of the way the ear responds to sound waves, "tones" at the different frequencies are present in the combination of sounds we hear and, in some circumstances, are easily distinguished.

TRY IT YOURSELF–TAKING IT FURTHER

1. Power varies as A^2. All other parameters in the power expression are constant. Let P be the power of a single wave with a 1.00-cm amplitude. Then $2P$ is the power of two waves not interfering. A single wave with twice the amplitude would have a power of $4P$. The wave in this problem has a power of $1.44^2 P = 2.07P$, more than the power of a single wave, and also more than the power of the two waves if they did not interfere!

2. At this wavelength the path difference will be 2.25 wavelengths, which is not a half-wavelength multiple. So for part (a) the intensity level will increase because complete destructive interference will not occur, and for part (b) the intensity level will decrease because complete constructive interference will no longer occur.

3. You would always have a ratio of two odd numbers, like 5:3 or 7:5.

4. Stringed instruments tend to go flat over time. The lower energy state is represented by a loss in tension, not an increase.

QUIZ

1. True

2. False

3. The ratio is about 2:1 because the length of the pipe is a half wavelength when both ends are closed and a quarter-wavelength when one end is open. It's not exactly 2:1 because the displacement antinode is not exactly at the open end but a little outside, so the effective air column is a little longer than the actual pipe. With both ends open, the fundamental is around 240 Hz again. (In fact it would be a little less than 240 Hz—perhaps 232 Hz—because the ends are open.) The fundamental frequency depends not on whether the ends are open or closed, but more on whether the two ends are the same or different.

4. All the frequencies that the clarinet produces are higher by the ratio of the speed of sound in helium to that in air, which is about 2.5 times. Thus all its tones sound more than an octave higher—provided that the clarinetist tightens his lips.

5. The phenomenon of beats is a clearly audible indicator of a small frequency difference. For example, when you simultaneously sound two strings that are supposed to have the same frequency, any difference in frequency produces beats that you can hear. Then you adjust the tension in one string, decreasing the beat frequency until no beats are heard.

6. 4.27 g

7. (a) $v = 342$ m/s, (b) $f = 171$ Hz

Chapter 17

COMMON PITFALLS

1. True
2. Celsius thermometers must agree that the ice point and boiling point of water are 0°C and 100°C, respectively, but they do not have to be in agreement for any other temperatures because not all thermometric properties vary with temperature in the same way.
3. True
4. On the Celsius scale, the temperatures that we ordinarily deal with tend to be numbers between -15°C and $+40$°C. These are more convenient to use and remember than temperature values of several hundred kelvins would be. The Kelvin scale is more suitable for scientific use in that many formulas are less complex when the Kelvin scale is used. Also, the Kelvin scale is independent of the thermometric properties of any particular substance.
5. True
6. The molar mass of CO is $12.0 + 16.0 = 28.0$, so a mole of CO has a mass of 28.0 g. There are Avogadro's number of molecules in a mole of any material.
7. False. Despite the relatively low density of molecules in our gaseous atmosphere (compared to liquids and solids), a molecule on average still has several collisions before it travels a distance of even one micrometer.
8. Yes. According to the equipartition theorem, there is an average energy of $\frac{1}{2}kT$ for each degree of freedom. So in addition to the $\frac{3}{2}kT$ average energy for translational motion, an O_2 molecule, for instance has two more degrees of freedom, adding another kT to the average energy. This is because the two oxygen molecules vibrate like two masses on opposite ends of a spring.

TRY IT YOURSELF–TAKING IT FURTHER

1. Yes; the temperature as defined by any thermometer we have talked about would vary linearly with the thermometric property, because by definition we measure the quantity at the boiling and freezing points of water and divide that difference into equally spaced quantities. However, if two thermometers that use different thermometric properties measure the same object at some other temperature, they will most likely *not* give the same temperature value of that object.
2. Driving the car increases the temperature of the tires and the air in them, which increases the pressure. The recommended tire pressures from the manufacturer already take this effect into account. If you measure the pressure and fill your tires when they are "hot" you will be under-inflating them.
3. Since the volume is constant, the pressure will be reduced proportionally to the reduction in temperature. The molecules can't go anywhere, however, so the chamber will still hold the same number of molecules. The next section explains how the same number of molecules can exert different pressures at different temperatures.
4. This method can certainly help narrow down the possible composition of a gas, but because there are several possibilities for each molar mass, additional testing is required.
5. Although this difference in average speed is small, in principle it could be used to separate gases made from different isotopes, possibly using something like the velocity selector discussed in Chapter 26.
6. The hydrogen doesn't drift off into space because only a very small fraction of it will have a great enough speed to escape the Sun's gravitational potential well.

QUIZ

1. True
2. False
3. The average molecular translational kinetic energy, which is proportional to the rms speed squared, must be reduced to $\frac{1}{4}$ its original value in order to halve the rms speed. Because this energy is proportional to the absolute temperature for an ideal gas, the absolute temperature must be reduced to $\frac{1}{4}$ its original value.
4. It depends a lot on what the thermometer is supposed to be good for, but there are some general requirements that apply to most cases. The thermometric property used—length, electrical resistance, pressure, or whatever—should be easily measurable itself and should vary linearly, or nearly so, with absolute temperature. Otherwise, the thermometer will be usable only for a limited number of applications. In most cases, the thermometer will need to be physically small so that it will quickly come to thermal equilibrium with the system to be measured, and also so that it will affect that system as little as possible. Other considerations will arise depending on the particular situation.

5. If two ideal gases are at the same temperature, they have the same average translational kinetic energy per molecule. If they do not happen to have the same molecular mass, however, then the molecules of the lighter gas will, on average, be moving at higher speeds.
6. (a) $K = 152$ J. (b) The kinetic energy will not change.
7. 236 atm

Chapter 18

COMMON PITFALLS

1. False. Doing work on a system can also cause an increase in its internal energy.
2. If the volume of the water is kept the same during the temperature increase (which is very difficult to do, by the way), then the water does no work.
3. False. The removal of heat could result in condensation or freezing of the material. These changes of phase do not decrease the temperature of the material.
4. Consider a process in which heat is added to a system and the system does work on its surroundings. The internal energy of the system will increase if the heat added to the system is greater than the work done by the system; the internal energy will remain the same if the heat added to the system is equal to the work done by the system, and the internal energy will decrease if the heat added to the system is less than the work done by the system. We can deduce these results from what we already know about work and energy. These results are encompassed by the first law of thermodynamics, which we will discuss further in the next section. Thus, it is possible for a system to absorb heat without its temperature and its internal energy increasing, if the work done by the system equals or is greater than the heat it absorbs.
5. False. If the heat lost by a system is greater than the work done on the system, then the system's internal energy will actually decrease.
6. Consider an ideal gas, confined in a cylinder with insulated walls, being compressed by a piston. The piston does work on the gas, which results in an increase in the internal energy of the gas. The internal energy of the gas is proportional to its temperature, so the increase in internal energy will result in an increase in temperature.
7. True
8. The internal energy of an ideal gas depends only upon its temperature. Thus in any isothermal process the internal energy of the gas remains constant. When a gas expands, it does work on its surroundings. According to the first law of thermodynamics, when the gas does work on its surroundings, its internal energy decreases by an amount equal to the work done, unless energy is transferred to the gas via heat. To keep the temperature—and thus the internal energy—constant, the heat added to the gas during expansion must equal the work done by the gas.
9. True
10. An ideal gas is initially at pressure P_0 and volume V_0. If the gas expands to a volume of $2V_0$ at constant pressure, the work it does on its surroundings is $P_0 \Delta V = P_0 V_0$. When a gas expands at constant temperature, the pressure varies inversely with the volume, so the pressure steadily drops as it expands. Thus, during the isothermal expansion the average pressure is always less than P_0, so the work done by the gas is less than $P_0 V_0$.
11. False. It is the other way around. The specific heat is the heat capacity per unit mass.
12. The internal energies per mole of all monatomic and diatomic gases are $1.5RT$ and $2.5RT$, respectively. Helium is a monatomic gas and nitrogen is diatomic. Thus at a given temperature the internal energy per mole for nitrogen is greater than it is for helium. The internal energy per gram of a substance equals the internal energy per mole divided by the molar mass, and the molar masses are 28 g/mol for nitrogen and 4 g/mol for helium. Thus, the internal energy per gram is $(2.5/28)RT = 0.089RT$ for nitrogen and $(1.5/4)RT = 0.375RT$ for helium. The internal energy per unit mass for helium is greater than it is for nitrogen.
13. False. In an adiabatic process, no heat is added to or removed from the system.
14. The ratio of molar heat capacities is $\gamma = 1 + R/c_V$ It is smaller when the number of degrees of freedom is larger because c_V is proportional to the number of degrees of freedom. A gas whose molecules have three or more atoms can be expected to have more degrees of freedom than a diatomic gas. Thus, for a gas with three or more atoms per molecule, the ratio of molar heat capacities is expected to be lower than it is for either monatomic or diatomic gases. Measurements show that many polyatomic gases have a ratio around $\gamma = 1.35$.

TRY IT YOURSELF–TAKING IT FURTHER

1. There are two reasons for this. First, there is more water than iron. But more significant is the relative

heat capacity of water compared to iron. It takes more than eight times as much heat to change the temperature of water by one degree than it does to change the temperature of iron by one degree.

2. More pans of hot water will be required. Heat losses to the tub and surroundings cannot be neglected in real life. This is especially true in this situation because it will likely take you at least 15–20 minutes to heat 8 pans of water, and that is a long time between pans in which the tub loses heat to the surroundings.

3. In that case the copper would not have enough heat to melt the ice, and the system would reach thermal equilibrium at a temperature of $0°C$, with some ice still present.

4. If the steam were at $150°C$, then no liquid water would be present. The steam could cool to some temperature above $100°C$ and provide the energy needed for all of the ice to turn into steam as long as the pressure in the flask remained sufficiently low.

5. Water is most dense at $4°C$, so its volume at this temperature will be less than its initial volume. That means the atmosphere will do positive work on the ice. In addition, the ice melts and is raised to a higher temperature. Each of these effects serves to increase the internal energy of the ice/water.

6. The question is really asking, which path requires the least heat? Based on the two paths given, path A requires less. However, a third path, in which the pressure is first dropped isovolumetrically, and then the volume is increased isobarically, will require even less heat, because the volume expansion of the gas does even less work.

7. In order to say much, you need to know what the nitrogen was replaced with. If it was replaced by He or Ne, for instance, which have smaller molar masses than N_2, you will have more than one-half mole of those gases, so it will take more heat to generate the same temperature change. However, for gases like Ar, Kr, or Xe, with heavier molar masses, you will have less than one-half mole, so less heat will be required.

8. The molar mass is approximately 28 g/mol, so the possibilities include CO, N_2, and C_2H_4.

9. It is most likely iron.

10. The final temperature is lower because no heat is added to or removed from the system, and the work done on the gas is negative.

11. The final temperature is higher because no heat is added to or removed from the system, and the work done on the gas is positive.

QUIZ

1. False

2. True

3. For any material that expands when its temperature is increased, c_p is greater than c_v, just as for a gas. That is so because, in an expansion at constant pressure, the material does work in pushing back its surroundings. We normally pay no attention to the distinction in solids and liquids because the change in volume is very small. The value ordinarily measured and tabulated is c_p.

4. Consider an ideal gas, confined in a cylinder with insulated walls, being compressed by a piston. The piston does work on the gas, which results in an increase in the internal energy of the gas. The internal energy of the gas is proportional to its temperature, so the increase in internal energy will result in an increase in temperature.

5. Isovolumetrically reduce the pressure to zero by removing heat from the system. Then isobarically reduce the volume. This will require no work because the pressure is zero. Then isovolumetrically increase the pressure by adding the heat you removed in the first step.

6. 0.132 kg

7. $30.1°C$

Chapter 19

COMMON PITFALLS

1. False. In an adiabatic compression no heat is transferred into or out of the system so the change in internal energy of the system is equal to the total work done on the system.

2. The force of kinetic friction always transforms mechanical energy into thermal energy. The engine is rated by how well it transfers mechanical energy to an external agent as work, so if some of the mechanical energy is dissipated via kinetic friction, the engine has less mechanical energy to transfer to the external agent. Thus the efficiency of the engine is reduced.

3. False. The COP (COP $= Q_c/W$) of a refrigerator is typically greater than 1.

4. Energy is transferred to the refrigerator's working substance both as heat from the things inside the box and as work done on it by the compressor. The refrigerator then transfers all of this energy as heat to the room air. With the refrigerator door kept open, the compressor has to work even harder.

This extra energy is transferred by the refrigerator to the room air as heat, causing the room to grow even hotter.

5. True. Any heat flow through a finite temperature difference is irreversible. Heat never flows from a lower temperature to a higher temperature.

6. This is consistent with the first law, which states that the heat absorbed by the gas equals the change in the internal energy of the gas plus the work done by the gas. The internal energy of an ideal gas depends only upon its temperature. Thus, in an isothermal expansion the change in internal energy is zero and the heat absorbed equals the work done.

7. False. $\Delta S = dQ_{rev}/T$ for a reversible process, but not all adiabatic processes are reversible. For example, an adiabatic free (unstrained) expansion of a gas is an irreversible process that results in an increase in entropy.

8. For the adiabatic, free expansion of an ideal gas no work is done by the gas. The first law tells us that the internal energy, and thus the temperature, do not change. Since the expansion is adiabatic, no heat is exchanged by the gas, which might tempt you to think that the change in entropy $\Delta S = \int dQ_{rev}/T$ is zero. However, this is incorrect. To calculate the change in entropy, we must consider a reversible process that brings the gas from its initial to its final state. Thus, we consider a reversible process that consists of two segments: a reversible adiabatic expansion to the final volume (during which work is done by the gas and so its temperature decreases), followed by a reversible, constant-volume warming where heat is absorbed by the gas until it reaches its final temperature. No heat is absorbed (or released) during the reversible adiabatic expansion. Therefore, during that segment the entropy of the gas does not change. During the reversible, constant-volume warming segment, heat is absorbed by the gas and its entropy increases. Therefore, in an adiabatic, free expansion, the entropy of an ideal gas must increase.

9. True

10. Let Q be the heat transferred and let T_1 be the higher and T_2 be the lower temperature. The change in entropy of the high-temperature object is $-Q/T_1$ and the change in entropy of the low-temperature object is $+Q/T_2$. Therefore, the change in entropy of the two objects, taken together, is $\Delta S = Q(1/T_2 - 1/T_1)$. If this is all that happens, then this is the total change in entropy of the universe. If, instead of just letting the heat flow, we had run a Carnot engine between these two objects as temperature reservoirs, the work we could have gotten from it is $W = \varepsilon_C Q = (1 - T_2/T_1)Q = T_2 \Delta S$. If, instead of running a Carnot engine, we just let the heat flow, this work would not have been done and the energy that could have done this work would no longer be available.

TRY IT YOURSELF–TAKING IT FURTHER

1. We must provide energy at a rate of $P_{out}/\varepsilon = 500$ W.

2. No. The COP is not a statement of conservation of energy. The work that must be done on the refrigerator is still the difference of the heat extracted from the cold reservoir and the heat rejected to the high-temperature reservoir. We cannot get something for nothing.

3. More water takes longer to freeze. If we use a higher-power refrigerator, then it will take less time to freeze the water. If the ambient air is warmer, then it will take longer to freeze the water.

4. No. The second law of thermodynamics says that no heat engine can achieve 100% efficiency.

5. Any number of things can contribute to a loss of efficiency. Friction is a significant source of lost efficiency; poor insulation is another.

6. A refrigerator must be used to extract heat from the water during freezing. As a consequence, the entropy of the high-temperature reservoir increases.

7. No. Changes in entropy depend only on the initial and final states, not on the process used to get from one state to the other.

8. The entropy of the surrounding air can decrease, as long as there is at least as great an increase in entropy somewhere else, like inside the freezer.

QUIZ

1. False

2. True

3. High efficiency is almost always a primary objective for a steam-electric generating plant. In such a plant, the steam is the working substance for the heat engine that drives the electric generator. The temperature of the steam is the temperature of the high-temperature heat source. The temperature of the low-temperature reservoir is usually fixed by circumstances, such as the temperature of a nearby lake. Thus, increasing the feed-steam temperature is the only way to increase the Carnot efficiency limit for the generator.

4. To say that a process is irreversible means, essentially, that the universe as a whole won't go in the other direction. In the examples given, we can put each "system" back where it started, but not

without making a permanent change in its "surroundings." That is, the agents used to return each "system" back to where it started are changed in the process, so the universe is not restored to its original state.

5. In any real process, the entropy of the universe will increase. In reality, if everything is at exactly the same temperature, it's all at thermal equilibrium and there can be no heat transfer; there must be some small temperature difference to have a heat transfer. However, if there is a temperature difference, the entropy loss Q/T_h of the warmer object is less than the entropy gain Q/T_c of the cooler object. The net entropy change of everything put together is always an increase.

6. a: $W = 3070$ J, $Q = 10,540$ J, $\Delta E_{int} = 7470$ J, $\Delta S = 24.4$ J/K; b: $W = 9870$ J, $\Delta E_{int} = 0$, $Q = 9870$ J, $\Delta S = 16.5$ J/K; c: $W = -3070$ J, $Q = -10,540$ J, $\Delta E_{int} = -7470$ J, $\Delta S = -24.4$ J/K; d: $W = -4940$ J, $\Delta E_{int} = 0$, $q = -4940$ J, $\Delta S = -16.5$ J/K; total: $W = 4930$ J, $Q = 4930$ J, $\Delta E_{int} = 0$, $\Delta S = 0$

7. (a) $\varepsilon = 0.380$, (b) $\Delta S = 0.288$ J/K, (c) $W_{lost} = 78.6$ J

Chapter 20

COMMON PITFALLS

1. False. Although most materials expand when heated, water contracts when it is heated to 4.00°C. Water is most dense at this temperature.

2. The hole enlarges with the plate. Every linear dimension of the metal plate undergoes the same fractional increase in length, including the diameter of the hole.

3. True

4. The van der Waals equation of state reduces to the ideal-gas law when the volume per mole is very large and therefore the density of the gas is very low.

5. True. The transfer of energy via radiation works this way, as does the conduction of heat.

6. The molecules of a gas spend most of their time between collisions. Thus, gases are good insulators with regard to the transfer of heat by conduction. They are poor insulators with regard to convection, however. By using materials that prevent convection currents from building up—materials that confine the gas in numerous small pockets—convective transfers of heat are all but eliminated. Thus, products used for insulation are light because they consist of numerous pockets of trapped gas. They occupy lots of space in order to minimize the temperature gradient, which in turn minimizes the thermal current.

TRY IT YOURSELF–TAKING IT FURTHER

1. The pin shrinks by roughly 100 μm. The width of a human hair is roughly 10 μm. So you might notice this much change if you looked carefully and it happened quickly. However, thermal cooling is a pretty slow process, so it's not likely you would see it in this case.

2. If the flask were filled with brass and then heated, the flask would break. Brass has a coefficient of expansion nearly six times that of Pyrex, so it will expand more. Since the Pyrex is not flexible, it will break under the increased pressure. Invar, however, has a smaller coefficient of expansion than Pyrex, so upon heating, the Invar will rattle around inside the flask, because it will not expand as much as the flask.

4. The greatest transfer of thermal energy is via conduction, so the most effective way to slow the vaporization is to minimize this. Replacing the steel supports with glass, which has a thermal conductivity roughly 10 times less than stainless steel should be quite effective.

5. The temperature of the junction will increase. First, the junction will be closer to the heat source. Secondly, copper is a better thermal conductor than aluminum, so it will more effectively transfer thermal energy from the hot bath to the junction than the aluminum did. If the short piece was aluminum and the long piece copper, then the answer would not be so clear.

QUIZ

1. False

2. No

3. If this happens, it is because the glass envelope holding the liquid gets hot first and expands a little before the temperature of the liquid inside has had time to rise. Then, as the glass, the liquid, and the surroundings all come to thermal equilibrium, the liquid rises with respect to the glass tube.

4. You may just be dissolving away sticky goo that is gluing the lid to the jar. More relevant to our discussion here is the fact that the metal lid will expand more as you increase its temperature than the glass jar will, so the lid loosens as the gap between the jar and the lid increases.

5. The vacuum between the double-glass walls prevents heat transfer through the walls by conduction and convection. Silvering the inner surfaces of the double walls cuts down on heat loss by radiation

by reducing their emissivity. The double walls must join at the neck, and there will be some heat loss by conduction there. This is reduced by using glass, which is a relatively poor thermal conductor.

6. 3.5 hours
7. 0.624 liters